NIST Special Publication 260-139

Standard Reference Materials®

Standard Reference Material 1750: Standard Platinum Resistance Thermometers, 13.8033 K to 429.7485 K

W. L. Tew
G. F. Strouse

Process Measurements Division
Chemical Science and Technology Laboratory
National Institute of Standards and Technology
Gaithersburg, MD 20899-8363

U.S. DEPARTMENT OF COMMERCE, *Donald L. Evans, Secretary*
TECHNOLOGY ADMINISTRATION, *Phillip J. Bond, Under Secretary for Technology*
NATIONAL INSTITUTE OF STANDARDS AND TECHNOLOGY, *Karen H. Brown, Acting Director*

Issued November 2001

Certain commercial equipment, instruments, or materials are identified in this paper in order to specify the experimental procedure adequately. Such identification is not intended to imply recommendation or endorsement by the National Institute of Standards and Technology, nor is it intended to imply that the materials or equipment identified are necessarily the best available for the purpose.

TABLE OF CONTENTS

PAGE

1. INTRODUCTION **2**
    1.1 Specifications 2
    1.2 Reference Functions 3
    1.3 Deviation Functions 4

2. CALIBRATION PROCEDURES **7**
    2.1 Fixed-Point Methods 7
    2.2 Comparison Methods 10

3. RESULTS **13**
    3.1 Deviations at Fixed-Point Temperatures 13
    3.2 Uncertainties 21
    3.3 Supplementary Data 25

4. SUMMARY **35**

5. REFERENCES **36**

APPENDIX: CERTIFICATE FOR SRM 1750 .... **39**

LIST OF FIGURES

FIGURE NO. PAGE

1. ITS-90 Reference Function for the range 13.8 K to 273.16 K.................................... 6
2. ITS-90 Reference Function for the range 273.15 K to 1234.93 K............................ 6
3. The immersion profile for a SRM 1750 SPRT in a NIST Hg TP cell...................... 8
4. The deviations for the SRM 1750 at various fixed-point temperatures.................. 15
5. The deviations for the SRM 1750 at the Hg TP vs. Ga MP................................. 15
6a. The deviations for the SRM 1750 at the Ar TP vs. Ga MP................................. 16
6b. The deviations for the SRM 1750 at the O_2 TP vs. Ga MP................................. 16
6c. The deviations for the SRM 1750 at the Ne TP vs. Ga MP................................. 17
6d. The deviations for the SRM 1750 at the e-H_2 VP_2 vs. Ga MP........................... 17
6e. The deviations for the SRM 1750 at the e-H_2 VP_1 vs. Ga MP........................... 18
6f. The deviations for the SRM 1750 at the e-H_2 TP vs. Ga MP............................. 18
7. The deviations of the lowest seven fixed points vs. the Ga MP expressed in mK....... 19
8. Individual deviation functions for the SRM 1750 vs. W value............................. 20
9. Individual deviation values for the SRM 1750 vs. temperature............................. 20
10. Uncertainty propagation curves for the range 13.8 K to 273.16 K....................... 23
11. Uncertainty propagation curve for the range 273.15 K to 429.7485 K.................. 24
12. Uncertainty propagation curve for the user's TPW measurement........................ 24
13a. The predicted and measured deviation differences for the first batch comparison.... 26
13b. The predicted and measured deviation differences for the second batch comparison 27
14. The correlation plot for the In FP vs. the Ga MP for the SRM 1750..................... 28
15a. The correlation plot for the Ar TP vs. the ^{4}He NBP.. 31
15b. The correlation plot for the O_2 TP vs. the ^{4}He NBP... 32
15c. The correlation plot for the Ne TP vs. the ^{4}He NBP... 32
15d. The correlation plot for the e-H_2 VP_2 vs. the ^{4}He NBP................................... 33
15e. The correlation plot for the e-H_2 VP_1 vs. the ^{4}He NBP................................... 33
15f. The correlation plot for the e-H_2 TP vs. the ^{4}He NBP..................................... 34

Standard Reference Material 1750: Standard Platinum Resistance Thermometers, 13.8033 K to 429.7485 K

Weston L. Tew and Gregory F. Strouse
National Institute of Standards and Technology
Process Measurements Division
Gaithersburg, MD 20899-8363

Abstract

The Standard Platinum Resistance Thermometer (SPRT) is defined by the International Temperature Scale of 1990 (ITS-90) as the interpolating instrument for temperatures between 13.8033 K and 1234.93 K. This SRM concerns the calibration characteristics of a group of 20 capsule-type SPRTs within the range of temperatures between 13.8033 K and 429.7485 K. The platinum wire used for the construction of the resistor elements of this SRM was derived from platinum bar stock with a chemical purity of 99.999 % by weight. Each of the 20 units has been evaluated, certified, and calibrated according to the definitions of the ITS-90. The resistance characteristics of all 20 units have been measured at 11 fixed point temperatures from 4.22 K to 429.7485 K as well as at least 7 other intermediate points. The procedures used in performing the calibrations and the supplemental measurements are documented in this publication. The resulting analysis pertaining to each individual unit, as well as the SRM sample population as a whole, is presented in detail.

Disclaimer

Certain commercial equipment, instruments, or materials are identified in this publication in order to adequately specify the experimental procedure. Such identification does not imply recommendation or endorsement by NIST, nor does it imply that these materials or equipment are necessarily the best available for the purpose.

Acknowledgement

The authors wish to thank the Standard Reference Materials Program for their support in the development of the SRM.

1. Introduction

This analysis concerns the 20 units of SRM 1750 Standard Platinum Resistance Thermometers (SPRTs) that have all been calibrated according to the definitions of the International Temperature Scale of 1990[1] (ITS-90). Each calibration consists of a series of resistance measurements of the SPRT at 10 different fixed-point temperatures. These fixed points are listed in Table 1, along with their defined temperatures, T_{90}, on the ITS-90.

Table 1. The 10 fixed points of the ITS-90 used for the SRM 1750 SPRT calibrations.

Notation	Descriptive Name	T_{90} / K
e-H_2 TP	equilibrium-hydrogen triple point	13.8033
e-H_2 VP_1	point of equilibrium-liquid hydrogen under a saturated vapor pressure near 33.3213 kPa	17.036
e-H_2 VP_2	point of equilibrium-liquid hydrogen under a saturated vapor pressure near 101.292 kPa	20.2714
Ne TP	neon triple point	24.5561
O_2 TP	oxygen triple point	54.3584
Ar TP	argon triple point	83.8058
Hg TP	mercury triple point	234.3156
H_2O TP	water triple point	273.16
Ga MP	gallium melting point	302.9146
In FP	indium freezing point	429.7485

The calibration data are used to produce a set of numerical coefficients for two deviation equations for the ITS-90 sub-ranges of 13.8 K to 273.16 K and 273.15 K to 429.7485 K. When these two deviation equations are combined with the appropriate ITS-90 reference functions, the SPRT can be used to interpolate all temperatures within the range of calibration. Further details concerning the structure of the scale may be found in NIST Technical Note 1265.[2]

1.1. Specifications

SPRT calibrations on the ITS-90 are parameterized in terms of the resistance ratio $W(T_{90})$ as defined by

$$W(T_{90}) \equiv \frac{R(T_{90})}{R(T_{H_2OTP})}, \tag{1}$$

where $R(T_{H_2OTP})$ is the SPRT resistance at the water triple point (H_2O TP). The most important specification for any SPRT is the ITS-90 purity criterion for the platinum element. This criterion is specified in terms of the measured $W(T_{90})$ value at the melting point of gallium (Ga MP) or the triple point of mercury (Hg TP), as given by

$$W(T_{GaMP}) \geq 1.11807, \tag{2}$$

or

$$W(T_{HgTP}) \leq 0.844235. \tag{3}$$

One of these inequalities must be met in order for an SPRT to serve as a defining instrument of the ITS-90. The platinum wire that was used in the construction of the SRM 1750 SPRTs was derived from bar stock with a purity of 99.999+ % as reported by the supplier[3]. However, the purity of the platinum is verified only for wire of 0.5 mm diameter or greater. Since the SRM 1750 SPRTs are constructed primarily with wire of 0.076 mm diameter, the actual purity level for that wire is expected to be slightly lower. Nonetheless, this platinum wire purity level is still sufficient to yield $W(T_{GaTP}) \geq 1.1181175$ and $W(T_{HgTP}) \leq 0.844165$ for SRM 1750, which well exceeds the ITS-90 criterion.

The resistance value for each of the SRM 1750 SPRTs at the H$_2$O TP is the customary value for capsule SPRTs, or $R(T_{H_2OTP}) = (25.5 \pm 0.1)$ Ω, but within a somewhat smaller tolerance than is normally found in commercial devices. Other important design specifications include a strain-free platinum element and a hermetically sealed capsule containing high-purity He gas. The design of the element used in the case of SRM 1750 is that of the so called "bird cage" type[4]. In these and other respects, the specifications for the SRM 1750 do not exceed, or significantly depart from, those that are customary for most commercial capsule SPRTs.[5]

The construction of the capsule for the SRM 1750 is typical of certain commercial types of SPRTs. The outer sheath is made from a nickel alloy with a 5.7 mm diameter which is lined with an inner platinum sheath. The overall length is approximately 46 mm and each unit is engraved on the outer sheath with the lettering "NIST SRM 1750," together with a four digit serial number (4450 through 4459 for the first calibration batch; and 4462, 4463, 4486 through 4493 for the second calibration batch). The four Pt lead wires are 0.3 mm in diameter and 10 cm long, insulated with a combination of polypyromelitimide (PPMI) and polytetraflouroethylene (PTFE), and terminated with 0.75 mm diameter gold-plated beryllium-copper contact pins. Each SRM 1750 SPRT is provided with 1.) a borosilicate glass adapter probe, 2.) a nylon tee connector for a purge gas connection and 3.) a polarized four-wire electrical connection with a lead-wire harness. The probes dimensions are 7.5 mm in outer diameter and approximately 60 cm in length.

1.2. Reference Functions

The ITS-90 specifies two reference functions that cover the overall SPRT temperature range, and we refer to them here as the lower range reference function and the upper range reference function. One reference function spans the range 13.8033 K to 273.16 K ("lower range"), and the other reference function spans the range 273.15 K to 1234.93 K ("upper range"). The reference functions, denoted as $W_r(T_{90})$, are representations based on two real SPRTs of very high purity and are designed to closely approximate any real SPRT resistance ratio, $W(T_{90})$. Figures 1 and 2 are plots of the lower range and upper range reference functions respectively.

The lower range reference function is given by

$$\ln[W_r(T_{90})] \equiv A_0 + \sum_{i=1}^{12} A_i \left[\frac{\ln(T_{90}/273.16\text{ K}) + 1.5}{1.5}\right]^i, \qquad (4)$$

where the values for the A_i are given in Table 2A and 13.8033 K $\leq T_{90} \leq$ 273.16 K.

The upper range reference function is given by

$$W_r(T_{90}) \equiv C_0 + \sum_{i=1}^{9} C_i \left[\frac{(T_{90}/\text{K}) + 754.15}{481}\right]^i, \qquad (5)$$

where the values for the C_i are given in Table 2B and 273.15 K $\leq T_{90} \leq$ 1234.93 K.

Table 2A. Lower range reference function.

A_0	-2.135 347 29×10^{+0}
A_1	3.183 247 20×10^{+0}
A_2	-1.801 435 97×10^{+0}
A_3	7.172 720 40×10^{-1}
A_4	5.034 402 70×10^{-1}
A_5	-6.189 939 50×10^{-1}
A_6	-5.332 322 00×10^{-2}
A_7	2.802 136 20×10^{-1}
A_8	1.071 522 40×10^{-1}
A_9	-2.930 286 50×10^{-1}
A_{10}	4.459 872 00×10^{-2}
A_{11}	1.186 863 20×10^{-1}
A_{12}	-5.248 134 00×10^{-2}

Table 2B. Upper range reference function.

C_0	2.781 572 54×10^{+0}
C_1	1.646 509 16×10^{+0}
C_2	-1.371 439 00×10^{-1}
C_3	-6.497 670 00×10^{-3}
C_4	-2.344 440 00×10^{-3}
C_5	5.118 680 00×10^{-3}
C_6	1.879 820 00×10^{-3}
C_7	-2.044 720 00×10^{-3}
C_8	-4.612 200 00×10^{-4}
C_9	4.572 400 00×10^{-4}

1.3. Deviation Functions

The extent to which an SPRT does not perfectly conform to the ideal representation given by the reference function is referred to as the deviation, or deviation function, which is defined as the difference between the observed resistance ratio and the reference function value, or

$$\Delta W(T_{90}) \equiv W(T_{90}) - W_r(T_{90}) \qquad (6)$$

The SPRT deviation is that part of its resistance-temperature relationship which is considered unique to each SPRT and must be included in the form of a specified deviation function as part of the calibration process for each SPRT. The form of the deviation depends on which of the different calibration sub-ranges is being used. In all cases there are one or more unspecified coefficients in the deviation function that are uniquely determined for the SPRT from the fixed-point calibration data.

1.3.1. Calibration Sub-range 13.8033 K to 273.16 K

The SPRT resistance is measured at the equilibrium hydrogen triple point (e-H$_2$ TP), the equilibrium hydrogen vapor pressure point (e-H$_2$ VP$_1$) near 33.32 kPa ($T_{90} \cong$ 17.035 K), the equilibrium hydrogen vapor pressure point (e-H$_2$ VP$_2$) near 101.29 kPa ($T_{90} \cong$ 20.27 K), the neon triple point (Ne TP), the oxygen triple point (O$_2$ TP), the argon triple point (Ar TP), the mercury triple point (Hg TP), and the H$_2$O TP. The deviation function for this calibration sub-range is given by

$$\Delta W_1(T_{90}) \equiv a_1[W(T_{90})-1] + b_1[W(T_{90})-1]^2 + \sum_{i=1}^{5} c_i \{\ln[W(T_{90})]\}^{i+2}. \qquad (7)$$

The coefficients a_1, b_1, c_1, c_2, c_3, c_4 and c_5 are determined by solving a system of seven simultaneous equations using the fixed-point calibration data. These data must include the results of measurements of $W(T_{\text{e-H}_2\text{TP}})$, $W(T_{\text{e-H}_2\text{VP}_1})$, $W(T_{\text{e-H}_2\text{VP}_2})$, $W(T_{\text{NeTP}})$, $W(T_{\text{O}_2\text{TP}})$, $W(T_{\text{ArTP}})$ and $W(T_{\text{HgTP}})$ for the SPRT. This sub-range covers the entire span of the lower range reference function and is the most complex of all the SPRT calibration sub-ranges.

1.3.2. Calibration Sub-range 273,15 K to 429.7485 K

The SPRT resistance is measured at the indium freezing point (In FP) and at the H$_2$O TP. The deviation function for this calibration sub-range is given by

$$\Delta W_{10}(T_{90}) \equiv a_{10}[W(T_{90}) - 1]. \tag{8}$$

The coefficient a_{10} is determined by solving the equation using the fixed-point calibration data. These data must include the results of measurements of $W(T_{\text{InFP}})$ for the SPRT. The gallium melting point (Ga MP) is a redundant point within the calibration sub-range that is used as a checkpoint for the calibration.

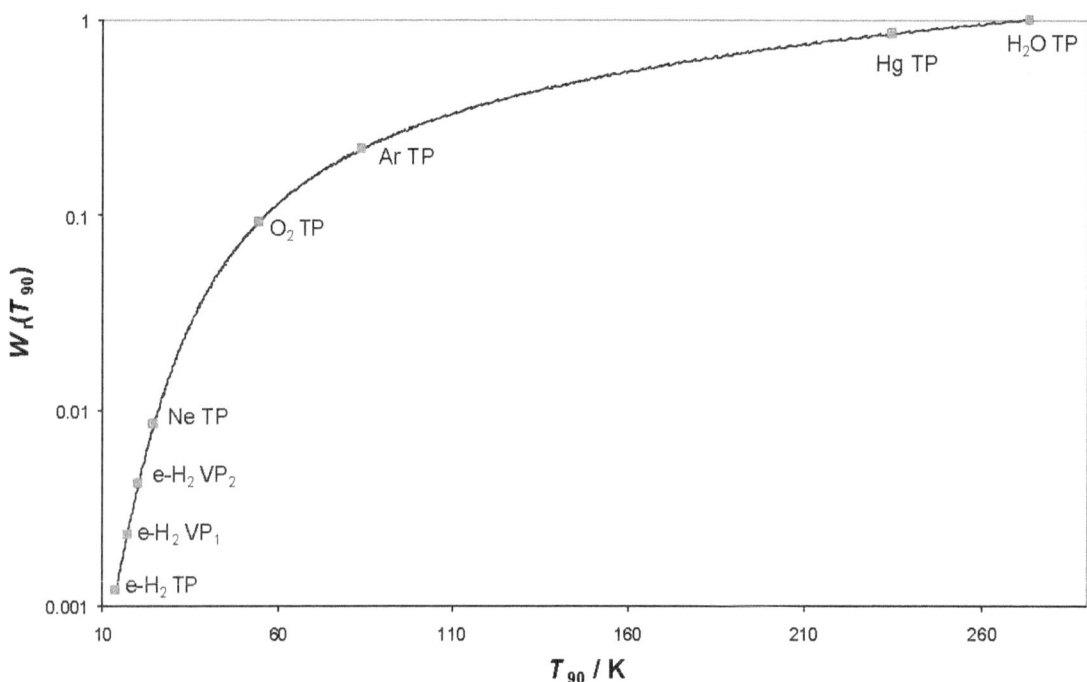

Figure 1. The SPRT reference function for the range 13.8 K to 273.16 K. The values of the function at the defining fixed-point temperatures are also shown.

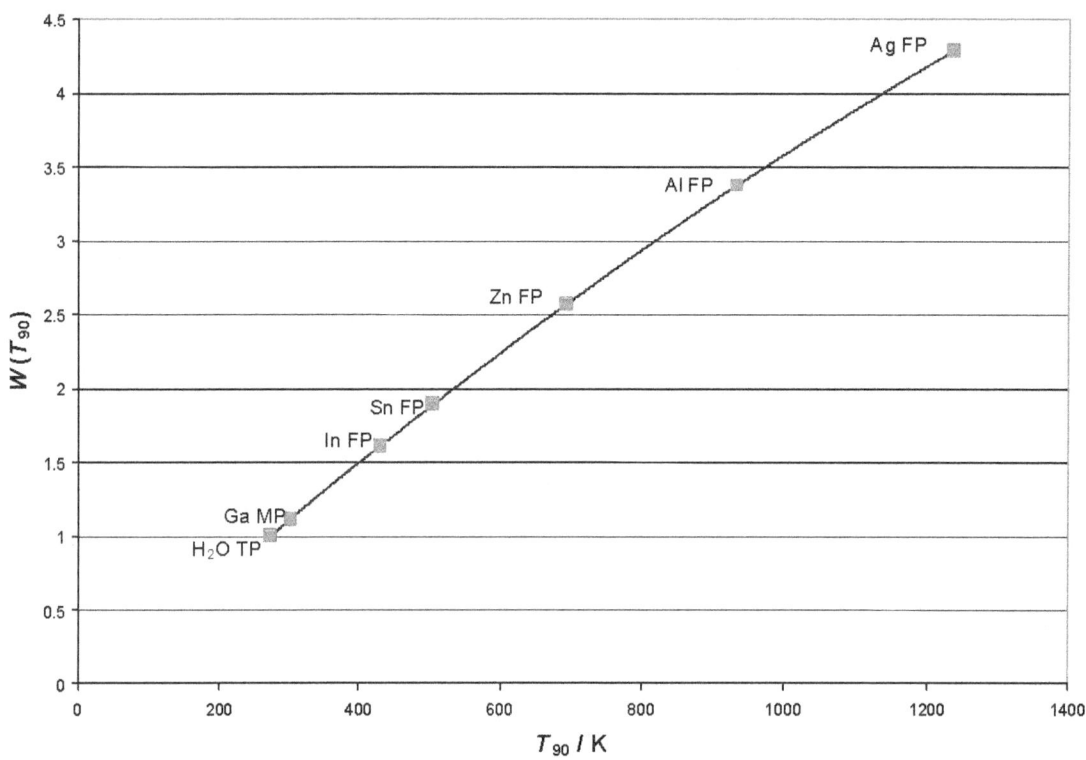

Figure 2. The SPRT reference function for the range 273.15 K to 1234.93 K. The values of the function at the defining fixed-point temperatures are also shown.

2. CALIBRATION PROCEDURES

As mentioned above, the SRM 1750 capsule SPRTs have been calibrated on the ITS-90 over the following two sub-ranges: 1) 13.8033 K to 273.16 K and 2) 273.15 K to 429.7485 K. This calibration involved the NIST realization of 10 different fixed points, although only nine are required (the Ga MP is redundant). Five of these fixed points (In FP, Ga MP, H$_2$O TP, Hg TP, and Ar TP) were realized directly with fixed-point cells at the time of the calibration of each SPRT.[6] The other five fixed points (e-H$_2$ TP, e-H$_2$ VP$_1$, e-H$_2$ VP$_2$, Ne TP, and O$_2$ TP) were realized at an earlier time using fixed-point cells and NIST reference capsule SPRTs.[7] The lower five fixed-point temperatures were then transferred to the SRM SPRTs via comparison to the reference SPRTs.[8] This is essentially the same method that is normally used at NIST to calibrate any customer SPRT over this same range of temperatures.

Measurements of each SPRT's resistance at the respective fixed-point temperatures were made via an automatically balancing ac resistance-ratio bridge. The ratio measurements were performed at a carrier frequency of 30 Hz, using either a 1 Ω, 10 Ω, or 100 Ω resistance standard, R_{std}, and the various excitation currents, i_1, i_2, as shown in Table 3. Other parameters shown in Table 3 are the values of the SPRT reference function $W_r(T_{90})$, and the approximate measured resistance ratio $r \equiv R(T_{90})/R_{std}$.

Table 3. Measurement parameters for the 10 fixed-point temperatures used for SRM 1750.

Fixed Point	T_{90} / K	$W_r(T_{90})$	r	R_{std} / Ω	i_1, i_2 / mA
In FP	429.7485	1.60980185	0.410	100	1.0, 1.414
Ga MP	302.9146	1.11813889	0.285	100	1.0, 1.414
H$_2$O TP	273.16	1.00000000	0.255	100	1.0, 1.414
Hg TP	234.3156	0.84414211	0.218	100	1.0, 1.414
Ar TP	83.8058	0.21585975	0.055, 0.55	100, 10	1.0, 2.0
O$_2$ TP	54.3584	0.09171804	0.234	10	1.0, 2.0
Ne TP	24.5561	0.008449736	0.215	1	2.0, 2.828
e-H$_2$ VP$_2$	20.2714	0.004235356	0.108	1	2.0. 2.828
e-H$_2$ VP$_1$	17.036	0.002296459	0.058	1	2.828, 5.0
e-H$_2$ TP	13.8033	0.001190068	0.030	1	2.828, 5.0

2.1. Fixed-Point Methods

The use of capsule SPRTs in fixed-point cells that are designed for long-stem thermometers requires an adapter probe for the capsule. A variety of such probes have been constructed in the past at NIST[9] and elsewhere[10] for this purpose. For SRM 1750, NIST has chosen a precision-bore borosilicate tube which closely fits the diameter of the SPRT. The probes are used to calibrate the SPRTs in the five fixed-point cells referred to above. When in use, the probes are pressurized with He gas to a level slightly above the ambient atmospheric pressure.

Any adapter probe constructed for SPRT calibration must achieve adequate immersion in the fixed-point cells.[11] This can be demonstrated by performing an immersion profile measurement in which measurements are taken as the probe is step-wise lowered into the last few centimeters of the cell's re-entrant well. If the immersion is adequate over this region, the data should produce the profile of the static pressure-head effect for the particular material in the cell.[12]

The immersion characteristics of an SRM 1750 SPRT in its glass adapter probe are shown in Figure 3 for a NIST Hg TP cell.[13] The data are shown for SPRT resistances extrapolated to zero power dissipation as a function of immersion depth relative to the bottom of the re-entrant well. The theoretical immersion profile is shown according to the static pressure head generated by the mercury, which is 7.1 mK/m or about 7.1 µΩ/cm. The data are in good agreement with the expected profile and indicates the immersion of this probe is adequate for this Hg TP cell.

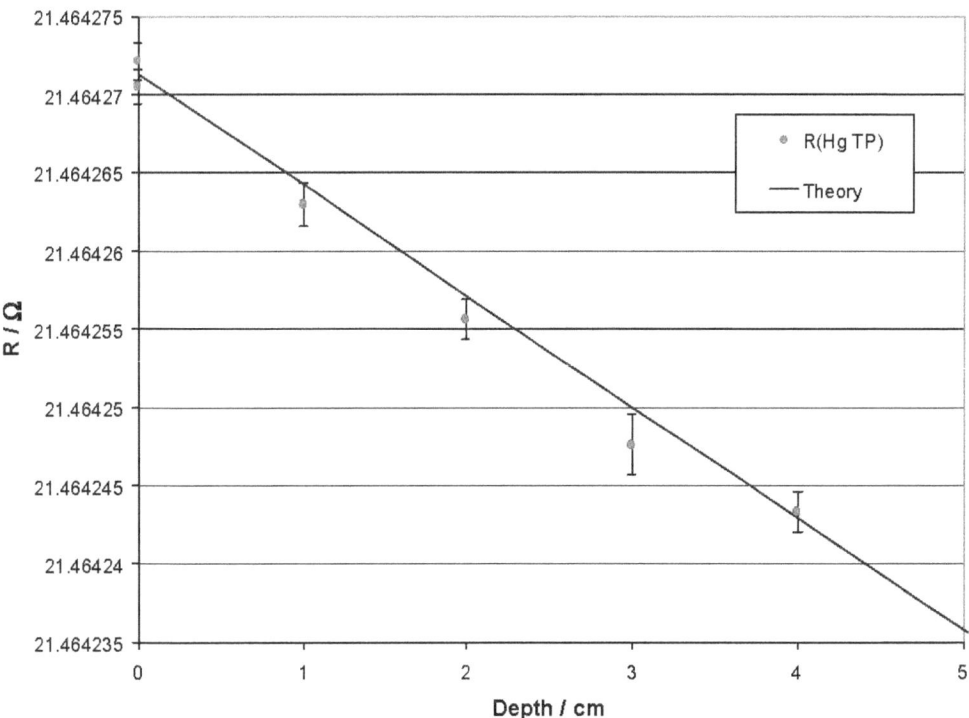

Figure 3. The immersion profile for a SRM 1750 SPRT in a NIST Hg TP cell. Full immersion is 15 cm between the element center and the top of the solid/liquid metal column, shown as "0 cm" depth in the figure. The error bars represent type A $k=1$ uncertainties.

The standard calibration process control employed at NIST for any SPRT calibration using fixed-point cells involves the use of NIST check thermometers with every fixed point. For the fixed-point cells discussed here, the check thermometers are long-stem SPRTs, each of which is dedicated for use with a particular fixed point.

2.1.1. The freezing point of indium

The serial number of the In freezing point cell used in the realizations for SRM 1750 is In 96-3. The metal purity is 99.999 99 % as derived from SRM 1745.[14] The static pressure-head correction applied was 0.59 mK based on an immersion depth of 18 cm from SPRT element midpoint to the top surface of the metal column.

2.1.2. The melting point of gallium

The gallium melting point was realized by a melting plateau under saturated vapor pressure conditions, (i.e., as a triple point). A correction of 2.01 mK was applied to correct for the absence of an external pressure. The serial number of the Ga triple-point cell used in the realizations for SRM 1750 is Ga 943. The metal purity is 99.999 995 % as derived from recent NIST results[15] in preparation for a future SRM. The static pressure-head correction applied was −0.22 mK based on an immersion depth of 18 cm from the SPRT element midpoint to the top surface of the metal column.

2.1.3. The triple point of water

The serial numbers of the water triple point cells used in the realizations for SRM 1750 are A-13-1126, A-13-1287, and A-13-1288. These cells are essentially indistinguishable in their characteristics and were all made and purified at a local commercial facility[16] from highly distilled water derived from the Potomac River. The exact purity of the water in these cells is unknown, however, they have been compared with other water triple point cells from other national laboratories and are generally accepted as the highest quality available today[17]. The static pressure-head correction applied was −0.186 mK based on an immersion depth of 26.5 cm from the SPRT element midpoint to the top surface of the water column.

2.1.4. The triple point of mercury

The serial number of the Hg triple point cell used in the realizations for SRM 1750 is Hg SS-1.[18] The metal purity is 99.999 999 % as derived from SRM 743.[19] The static pressure head correction applied was 1.065 mK based on an immersion depth of 15 cm from the SPRT element midpoint to the top surface of the metal column (20 % melted fraction).

2.1.5. The triple point of argon

The argon triple-point cell used in the realizations for SRM 1750 is a large apparatus with six wells designed for long-stem SPRTs and one central well designed for capsule SPRTs[20]. The SRM 1750 SPRTs were all calibrated in this apparatus in the long-stem SPRT wells using the glass adapter probes. For the purposes of this report, the argon apparatus is designated as "Ar-NIST-1X7." The triple point is realized in a melting mode at approximately 20 % melted fraction. The gas purity of the argon as supplied from a commercial source is 99.9999 %.[21] The static pressure head correction applied was 0.36 mK based on an immersion depth of 10.9 cm from the SPRT element midpoint to the top surface of the solidified gas column (20 % melted fraction).

2.2. Comparison Methods

Comparison measurements for capsule SPRTs are performed at temperatures of 83.8 K and below. The comparisons for SRM 1750 were made in two batches of ten SPRTs each. The batch of SRM thermometers, along with a capsule reference thermometer and three other check thermometers, were placed in a series of close-fitting wells within a high-purity copper comparison block. The comparison block was controlled at an appropriate constant temperature under high vacuum conditions and surrounded by a series of isothermal shields.

Each SRM batch thermometer (BT) was measured at the two excitation currents i_1 and i_2 (see Table 3) in a symmetrical sequence, starting and ending with the reference thermometer (RT). The sequence was a series of seven measurement records: RT(i_1), RT(i_2), BT(i_1), BT(i_2), BT(i_1), RT(i_2), RT(i_1). Each measurement record in the sequence consists of an average of 20 individual bridge readings. All reference SPRT measurements were made with two currents and resistance ratios corrected to zero-power dissipation. The average zero-power resistance of the reference thermometer's initial and final records was then used to compute the block temperature for the particular batch thermometer being measured. The reference thermometer was a capsule SPRT (serial no. 1004131) previously calibrated at NIST according to the ITS-90[22] using recent realizations of all required fixed points. These fixed-point realizations were performed according to any one of three independent techniques used at NIST.[20,23,24] The check thermometers consisted of two other similarly calibrated capsule SPRTs and two calibrated Rhodium-Iron Resistance Thermometers (RIRTs). These check thermometers provided checks on the entire calibration process through comparison of their indicated temperatures.

2.2.1. The triple point of argon

This comparison point was equivalent to and, hence, redundant with the fixed-point realization described under section 2.1.5. It was performed primarily as a systematic check to verify the consistencies between the two separate measurement systems which were used. Comparison measurements at the Ar TP temperature with the first batch of SRM 1750 SPRTs were performed in October 1998. Comparison measurements at the Ar TP temperature with the second batch of SRM 1750 SPRTs were performed in April 1999. During the first batch comparison calibration, comparisons between the reference SPRT (s/n 1004131) and check SPRTs (s/n 1842385 and s/n 1812279) indicated agreement to within 0.2 mK in indicated temperature. During the second batch comparison calibration, comparisons between the reference SPRT (s/n 1004131) and check SPRTs (s/n 1812282 and s/n 1812284) indicated agreement to within 0.02 mK in indicated temperature. These levels of agreement are generally within the expanded uncertainty for the comparison process at 83.8058 K.

The reference SPRT 1004131 was originally calibrated at the argon triple point at NIST in December of 1994 according to the techniques given by Meyer and Reilly.[23] This reference SPRT has recently undergone additional check realizations at NIST with other argon triple point cells including Ar-NBS-1[25] and Ar-NIST-1X7. The argon gas used in these other cells was derived from the same source,[21] as was the case for all NIST realizations of the Ar TP. Those check realizations indicate consistency with the 1994 realization to within 0.2 mK.

2.2.2. The triple point of oxygen

Comparison measurements at the O_2 TP temperature were performed in October 1998 with the first batch of SRM 1750 SPRTs, and in April 1999 with the second batch. During the first batch comparison calibration, comparisons between the reference SPRT (s/n 1004131) and check SPRTs (s/n 1842385 and s/n 1812279) indicated agreement to within 0.1 mK in indicated temperature. During the second batch comparison calibration, comparisons between the reference SPRT (s/n 1004131) and check SPRTs (s/n 1812282 and s/n 1812284) indicated agreement to within 0.01 mK in indicated temperature. These levels of agreement are generally within the expanded uncertainty for the comparison process at 54.3584 K.

The reference SPRT 1004131 was originally calibrated at the O_2 TP at NIST in February 1995 according to the techniques given by Meyer and Reilly.[23] This reference SPRT has recently undergone additional check realizations at NIST with one other O_2 TP cell, "PO-1."[26] The oxygen gas used in this other cell was derived from the same source[26] as was used in the 1995 realization. Those check realizations indicate consistency with the 1995 realization to within 0.05 mK.

2.2.3. The triple point of neon

Comparison measurements at the Ne TP temperature were performed in November 1998 with the first batch of SRM 1750 SPRTs and in May 1999 for the second batch. During the first batch comparison calibration, comparisons between the reference SPRT (s/n 1004131), the check SPRTs (s/n 1842385 and s/n 1812279), and the check RIRTs (s/n B-174 and s/n B-168) indicated agreement to within 0.32 mK in indicated temperature. During the second batch comparison calibration, comparisons between the reference SPRT (s/n 1004131), the check SPRTs (s/n 1812282 and s/n 1812284), and the check RIRTs (s/n B-174 and s/n B-168) indicated agreement to within 0.2 mK in indicated temperature. These levels of agreement are generally within the expanded uncertainty for the comparison process at 24.5561 K.

The reference and check SPRTs (s/n 1004131 and s/n 1842385) as well as the check RIRTs (s/n B-174 and B-168) were originally calibrated at the Ne TP at NIST in March 1995 according to the techniques given by Meyer and Reilly.[23] Independent realizations of the Ne TP using sealed cells[24] have recently been made at NIST using RIRT B-174 which indicate agreement with the 1995 realizations to within 0.2 mK.

2.2.4. The point of equilibrium-liquid hydrogen under a saturated vapor pressure near 101.292 kPa

Comparison measurements at an e-H_2 VP_2 temperature of 20.2741 K were performed in November 1998 with the first batch of SRM 1750 SPRTs and in May 1999 with the second batch. During the first batch comparison calibration, comparisons between the reference SPRT (s/n 1004131), the check SPRTs (s/n 1842385 and s/n 1812279), and the check RIRTs (s/n B-174 and s/n B-168) indicated agreement to within 0.21 mK in indicated temperature. During the second batch comparison calibration, comparisons between the reference SPRT (s/n 1004131), the check SPRTs (s/n 1812282 and s/n 1812284), and the check RIRTs (s/n B-174 and s/n B-168) indicated agreement to within 0.13 mK in indicated temperature. These levels of agreement are generally within the expanded uncertainty for the comparison process at 20.274 K.

The reference and check SPRTs (s/n 1004131 and s/n 1842385) were originally calibrated by comparison to a reference RIRT (s/n B-168) at 20.2693 K in May 1996. The check RIRTs (s/n

B-174 and B-168) were originally calibrated at an e-H$_2$ VP$_2$ temperature of 20.2693 K at NIST in June 1994 according to the techniques given by Meyer and Reilly.[27] Additional realizations of the e-H$_2$ VP$_2$ were performed at NIST in February 1999 using the check SPRTs (s/n 1812284 and s/n 1812282).[27]

2.2.5. The point of equilibrium-liquid hydrogen under a saturated vapor pressure near 33.3213 kPa

Comparison measurements at an e-H$_2$ VP$_1$ temperature of 17.036 K were performed in November 1998 with the first batch of SRM 1750 SPRTs and in May 1999 with the second batch. During the first batch comparison calibration, comparisons between the reference SPRT (s/n 1004131), the check SPRTs (s/n 1842385 and s/n 1812279), and the check RIRTs (s/n B-174 and s/n B-168) indicated agreement to within 0.14 mK in indicated temperature. During the second batch comparison calibration, comparisons between the reference SPRT (s/n 1004131), the check SPRTs (s/n 1812282 and s/n 1812284), and the check RIRTs (s/n B-174 and s/n B-168) indicated agreement to within 0.19 mK in indicated temperature. These levels of agreement are generally within the expanded uncertainty for the comparison process at 17.036 K.

The reference and check SPRTs (s/n 1004131 and s/n 1842385) were originally calibrated by comparison to a reference RIRT (s/n B-168) at 17.0352 K in May 1996. The check RIRTs (s/n B-174 and B-168) were originally calibrated at an e-H$_2$VP$_1$ temperature of 17.0352 K at NIST in June 1994 according to the techniques given by Meyer and Reilly.[27] Additional realizations of the e-H$_2$ VP$_2$ were performed at NIST in February 1999 using the check SPRTs (s/n 1812284 and s/n 1812282).[27]

2.2.6. The triple point of equilibrium-hydrogen

Comparison measurements at the e-H$_2$ TP temperature were performed in November 1998 with the first batch of SRM 1750 SPRTs and in May 1999 with the second batch. During the first batch comparison calibration, comparisons between the reference SPRT (s/n 1004131), the check SPRTs (s/n 1842385 and s/n 1812279), and the check RIRTs (s/n B-174 and s/n B-168) indicated agreement to within 0.15 mK in indicated temperature. During the second batch comparison calibration, comparisons between the reference SPRT (s/n 1004131), the check SPRTs (s/n 1812282 and s/n 1812284), and the check RIRTs (s/n B-174 and s/n B-168) indicated agreement to within 0.18 mK in indicated temperature. These levels of agreement are generally within the expanded uncertainty for the comparison process at 13.8033 K.

The reference and check SPRTs (s/n 1004131 and s/n 1842385) as well as the check RIRTs (s/n B-174 and B-168) were originally calibrated at the e-H$_2$ TP at NIST in April 1995 according to the techniques given by Meyer and Reilly.[23] Additional realizations of the e-H$_2$ TP were performed at NIST in December 1997 using the check SPRTs (s/n 1812284 and s/n 1812282).[27]

3. RESULTS

3.1. Deviations at Fixed-Point Temperatures

All results shown here are in terms of the individual deviations $\Delta W(T_{90})$ for all the SRM 1750 SPRTs that have been calibrated. A summary of the results for all 20 units is provided in Table 4. Only the zero-power values are tabulated here. The distribution of values is not at all Gaussian in shape; therefore the statistical characterization presented here is simply in terms of the maximum and minimum values ("max." and "min." in Table 4) of the deviations at each fixed-point temperature.

Table 4. Summary of $\Delta W(T_{90})$ values in for all 20 SPRTs of the SRM 1750.

Serial no.	In FP $\Delta W / 10^{-5}$	Ga MP $\Delta W / 10^{-5}$	Hg TP $\Delta W / 10^{-5}$	Ar TP $\Delta W / 10^{-5}$	O_2 TP $\Delta W / 10^{-5}$	Ne TP $\Delta W / 10^{-5}$	e-H_2 VP_2 $\Delta W / 10^{-5}$	e-H_2 VP_1 $\Delta W / 10^{-5}$	e-H_2 TP $\Delta W / 10^{-5}$
4450	-8.590	-1.697	1.686	9.446	10.928	11.188	10.762	10.249	9.513
4451	-10.424	-2.056	2.162	11.824	13.746	14.196	13.715	13.122	12.288
4452	-10.472	-2.035	2.159	11.618	13.484	13.999	13.572	13.005	12.137
4453	-9.291	-1.795	1.907	10.418	11.949	12.180	11.742	11.194	10.413
4454	-10.703	-2.052	2.226	11.689	13.609	14.088	13.614	13.015	12.121
4455	-9.114	-1.762	1.835	9.668	11.314	11.691	11.295	10.785	10.029
4456	-10.007	-1.944	2.067	10.843	12.547	12.761	12.269	11.649	10.765
4457	-10.687	-2.049	2.258	11.761	13.664	14.176	13.762	13.195	12.312
4458	-10.447	-2.035	2.184	11.476	13.424	13.571	12.922	12.187	11.203
4459	-10.949	-2.130	2.277	12.051	14.099	14.816	14.441	13.924	13.096
4462	-10.493	-2.073	2.272	11.808	13.550	13.743	13.243	12.653	11.790
4463	-8.040	-1.603	1.705	8.894	10.253	10.449	10.055	9.586	8.906
4486	-10.542	-2.010	2.228	11.513	13.390	14.035	13.636	13.102	12.255
4487	-10.311	-1.994	2.167	11.238	13.028	13.552	13.114	12.561	11.719
4488	-10.619	-2.064	2.232	11.545	13.549	14.278	13.912	13.400	12.565
4489	-11.037	-2.111	2.217	12.178	14.191	14.636	14.164	13.574	12.674
4490	-11.496	-2.238	2.413	12.562	14.685	15.298	14.826	14.218	13.286
4491	-11.185	-2.151	2.317	12.241	14.282	14.822	14.338	13.721	12.790
4492	-11.407	-2.174	2.378	12.591	14.727	15.307	14.839	14.228	13.280
4493	-10.858	-2.110	2.288	11.779	13.816	14.455	14.032	13.469	12.587
min.	-11.496	-1.603	1.686	8.894	10.253	10.449	10.055	9.586	8.906
max.	-8.040	-2.238	2.413	12.591	14.727	15.307	14.839	14.228	13.286

3.1.1. Correlations in the Fixed-Point Deviation Data

The same data contained in Table 4 are shown graphically in Figure 4, except that the e-H_2VP_1 and e-H_2VP_2 data have been omitted for clarity. Here the $\Delta W(T_{GaMP})$ values are plotted on the abscissa, and all other $\Delta W(T_{90})$ values are plotted on the ordinate. The origin of the plot represents values equivalent to the reference function. The linear trends that are shown in Figure 4 have ordinate intercepts that are near this origin. This feature would be expected for platinum of comparable purity to that used in the SPRTs upon which the reference functions are based. The high degree of correlation in the deviations between all of the fixed points is also evident. This correlation is an indication of the uniformity in the starting material, in this case the platinum bars used to draw the wire for the SPRT elements, as well as the uniformity of the annealing process used for the elements after they have been formed. Furthermore, a linear correlation should always be expected for high-purity-metal resistivity data of this type as long as the basic mechanisms responsible for the electron scattering obey, or at least approximate, Matthiessen's rule. Similar linear relationships in correlation plots have been found in historical studies[28] of high purity platinum and modeled as specific forms of deviations from Matthiessen's rule.

The correlation plot of the $\Delta W(T_{HgTP})$ values versus the $\Delta W(T_{GaMP})$ values is shown in Figure 5 on two different scales. The larger-scale plot is illustrative of the purity of the SRM 1750 platinum relative to the ITS-90 criterion for the W values of those two fixed points. In general, all SPRTs suitable for the ITS-90 will have deviations on a plot of this type which lie within a diagonal band approximately 1×10^{-5} (2.5 mK) in width running between the origin and the intersection of the criterion lines. The fact that the SRM 1750 SPRTs are distributed relatively close together and closer to the origin (reference function) is again a measure of their relatively high purity and uniformity of platinum. The fine-scale insert plot of figure 5 shows the same data with NIST estimated uncertainties[29,30] as confidence-limit bars (see section 3.2 below).

The six lowest temperature fixed-point deviations are shown in Figures 6a through 6f, again plotted as correlations with the $\Delta W(T_{GaMP})$ deviations. The linear least-squares fits are included to help clarify the general trend of the data in the presence of small amounts of sample-to-sample variation. Despite this sample dependent variation, the overall limits of the deviation distributions (maximum-minimum) remains relatively constant among these lower temperature fixed points. The linear fit is shown in equation form on each plot. The size of the bounds on the data shown in each plot is $\pm 2\sigma_r$ where σ_r is the standard deviation in the residuals of each fit as expressed in equivalent temperature units.

It is illustrative to plot some of these same data already presented in terms of equivalent deviations in temperature rather than resistance ratio. Figure 7 is another deviation plot of $\Delta W(T_{90})$ values on a logarithmic scale versus $\Delta W(T_{GaMP})$ on a linear scale except that all deviations are now converted to temperature differences and expressed in mK. In this plot the deviations now appear more rationally ordered and are roughly proportional to the inverse temperature.

Finally, Table 5 lists a summary of all the statistical parameters derived from the correlation plots. The linear fits shown in the correlation plots are derived according to a least-squares method. These fits are parameterized according to the general linear fit formula

$$\Delta W(X) = m \Delta W(\text{Ga MP}) + \Delta_0 , \qquad (9)$$

where X represents the particular fixed point of interest. The slope m has an associated standard deviation σ_m, and the intercept Δ_0 has an associated standard deviation σ_Δ for each correlated pair of fixed points. In addition, the standard deviation σ_r of the residuals in $\Delta W(X)$ for each fit are tabulated both as dimensionless deviations ($/10^{-5}$) and as the temperature deviation equivalents in units of mK.

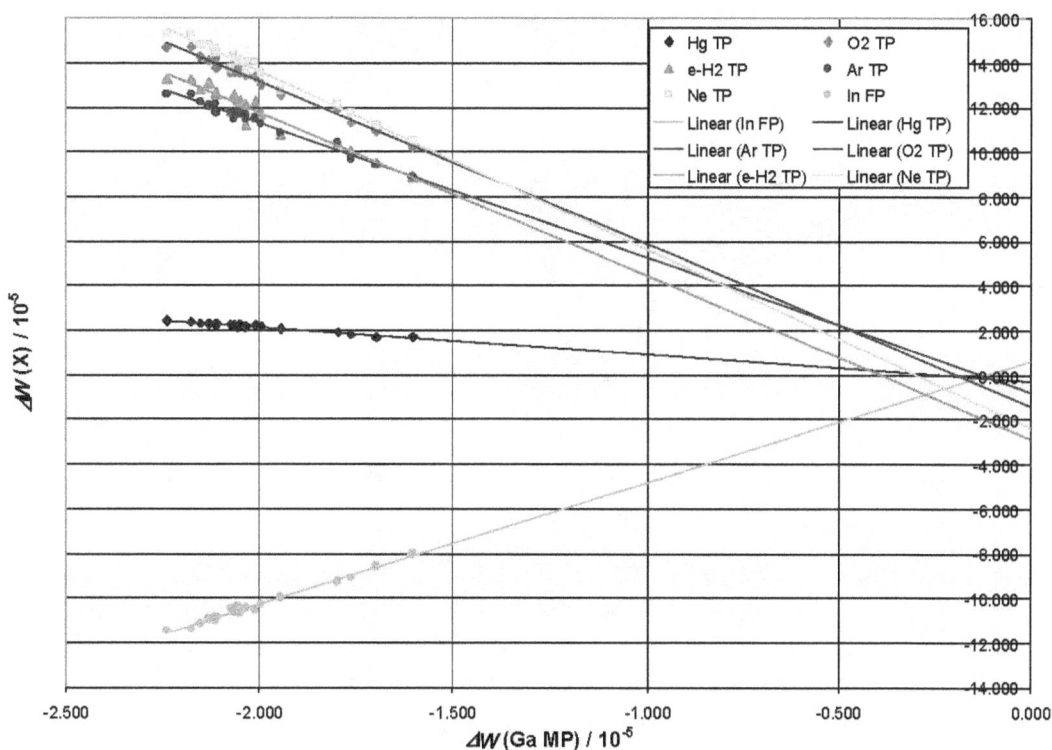

Figure 4. The deviations for the SRM 1750 population at the fixed-point temperatures for the e-H$_2$ TP, Ne TP, O$_2$ TP, Ar TP, Hg TP, and In FP versus that for the Ga MP.

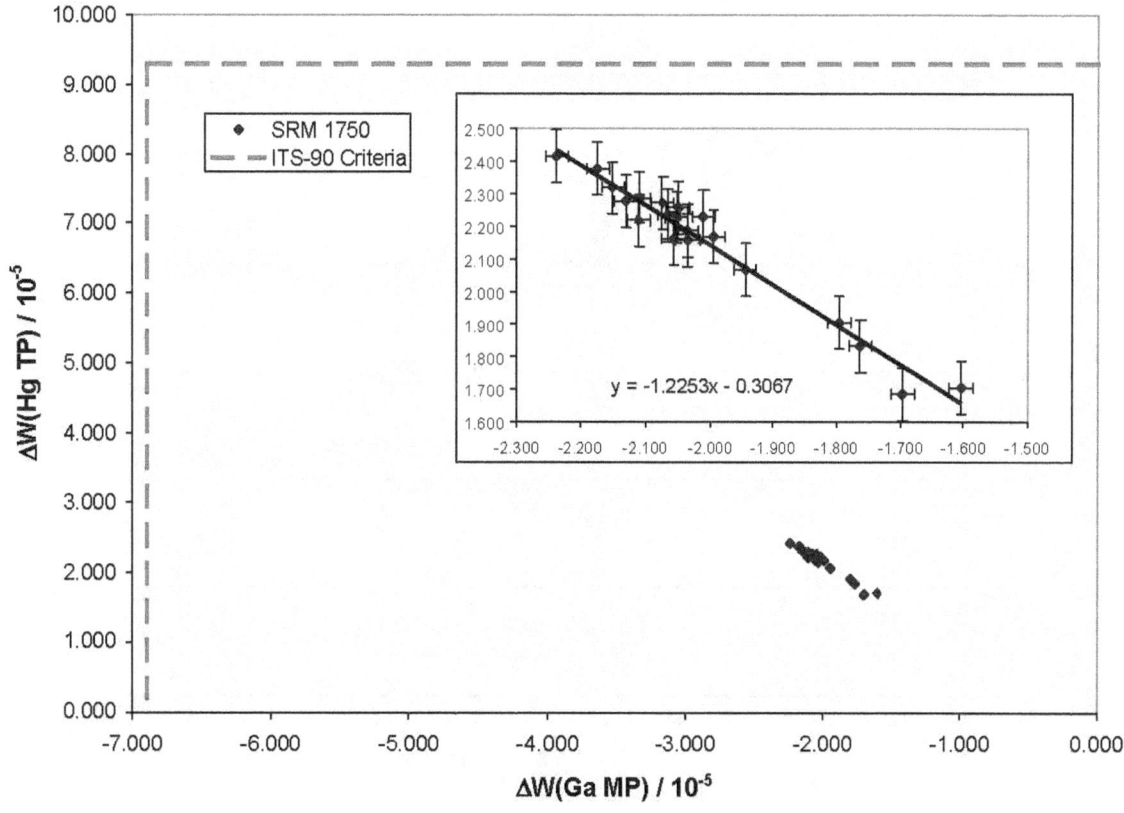

Figure 5. The deviations for SRM 1750 at the Hg TP versus the Ga MP shown on a scale which includes the cutoff limits specified by the ITS-90, "ITS-90 Criteria." The inset plot is the same data with a smaller scale.

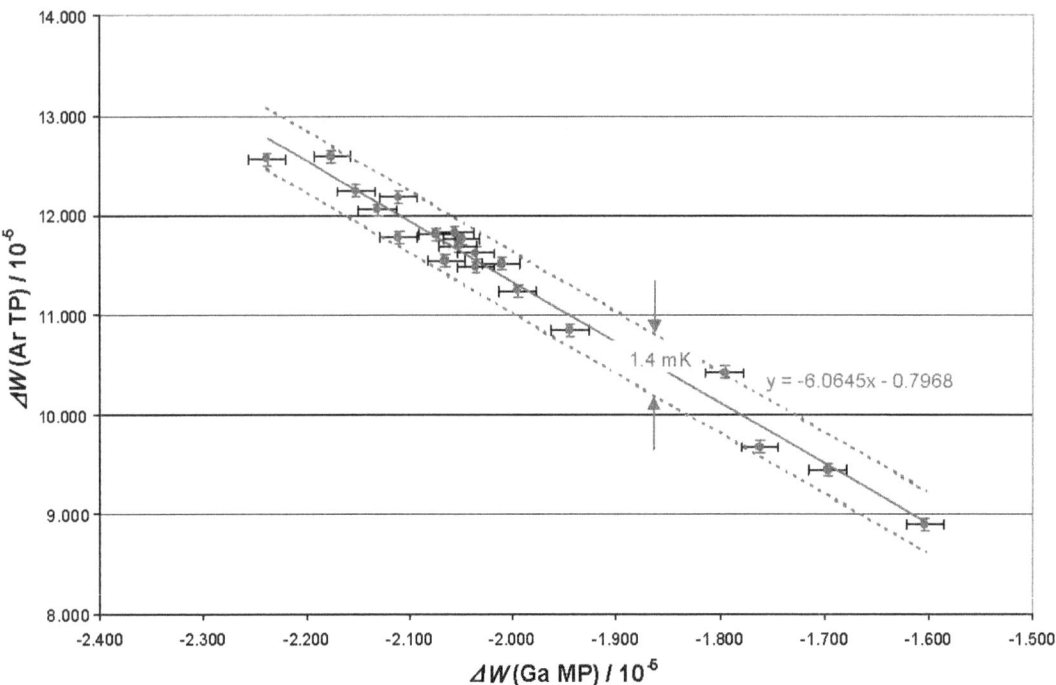

Figure 6a. The correlation plot for the deviations at the Ar TP versus the Ga MP. The central curve is a simple linear least-squares fit. The dotted lines are the symmetric bounds on the data about the linear fit with the size of the bound indicated as an equivalent temperature difference.

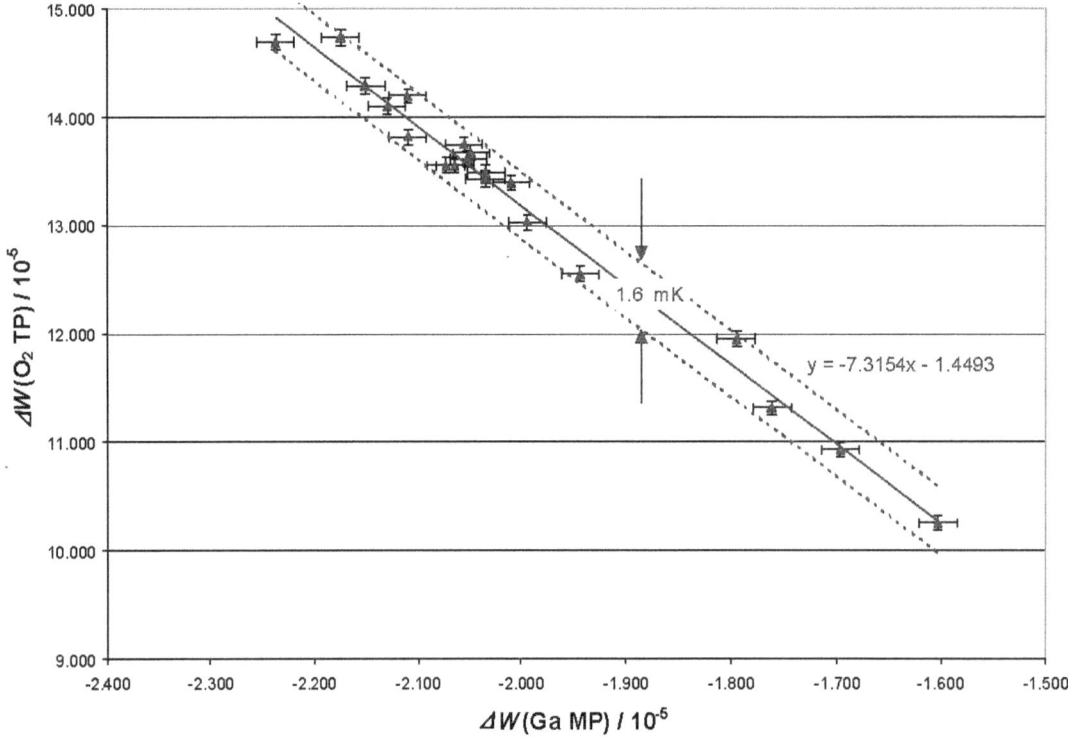

Figure 6b. The correlation plot for the deviations at the O_2 TP versus the Ga MP. The central curve is a simple linear least-squares fit. The dotted lines are the symmetric bounds on the data about the linear fit with the size of the bound indicated as an equivalent temperature difference.

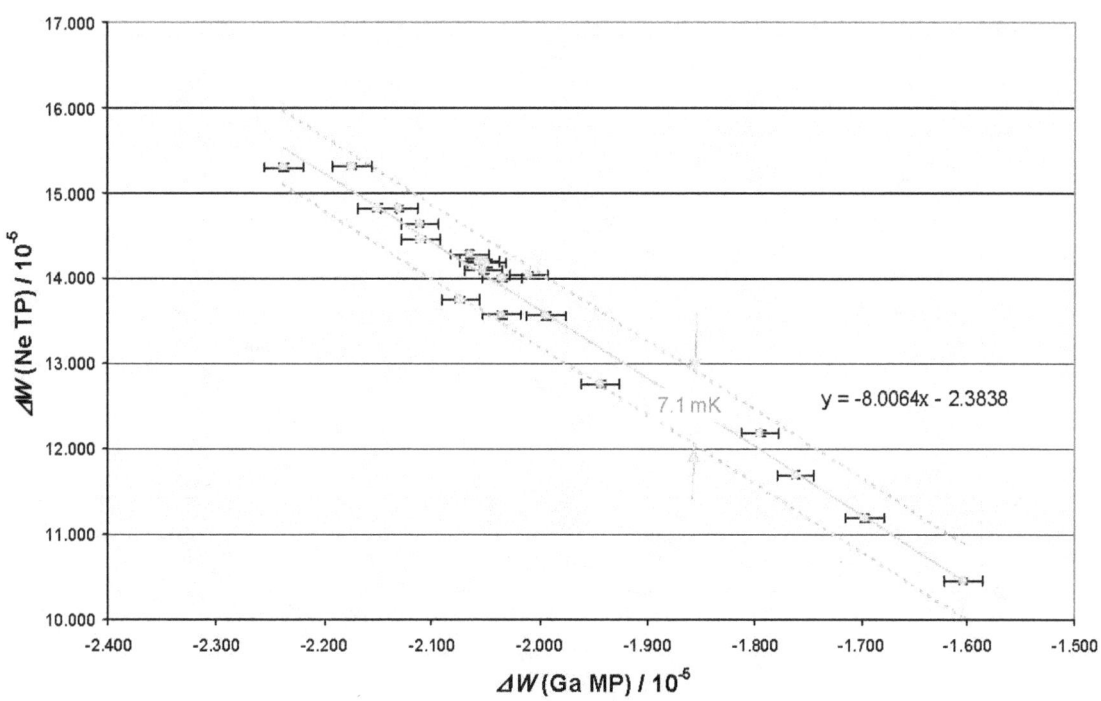

Figure 6c. The correlation plot for the deviations at the Ne TP versus the Ga MP. The central curve is a simple linear least-squares fit. The dotted lines are the symmetric bounds on the data about the linear fit with the size of the bound indicated as an equivalent temperature difference.

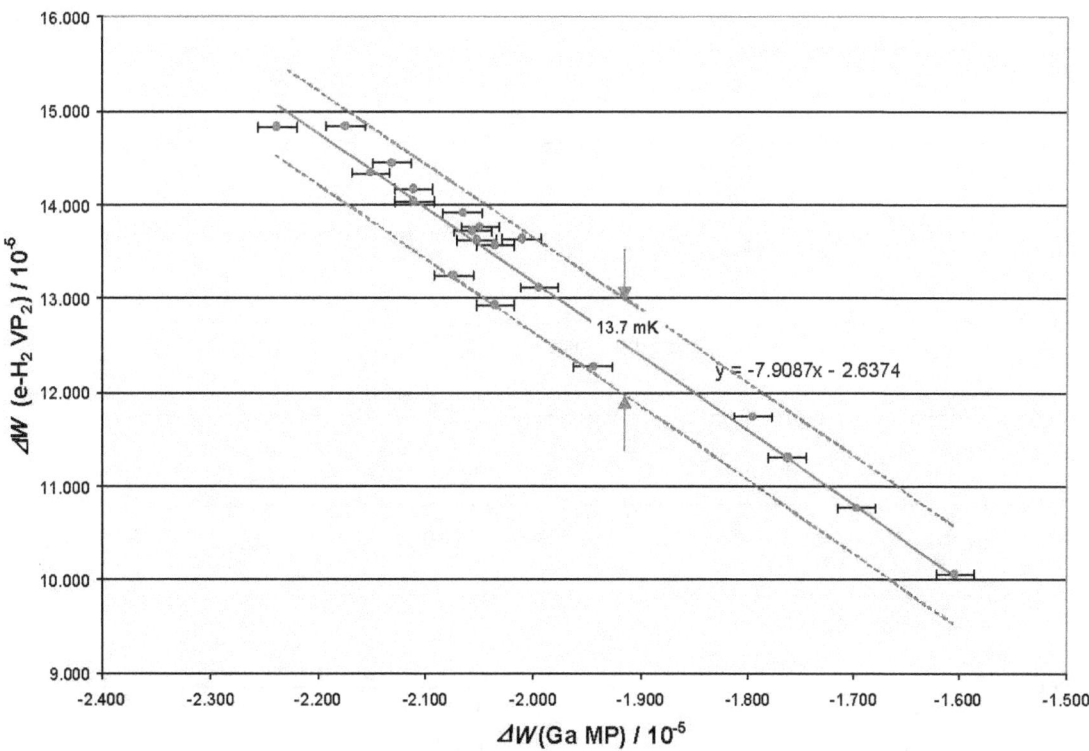

Figure 6d. The correlation plot for the deviations at the e-H_2 VP_2 versus the Ga MP. The central curve is a simple linear least-squares fit. The dotted lines are the symmetric bounds on the data about the linear fit with the size of the bound indicated as an equivalent temperature difference.

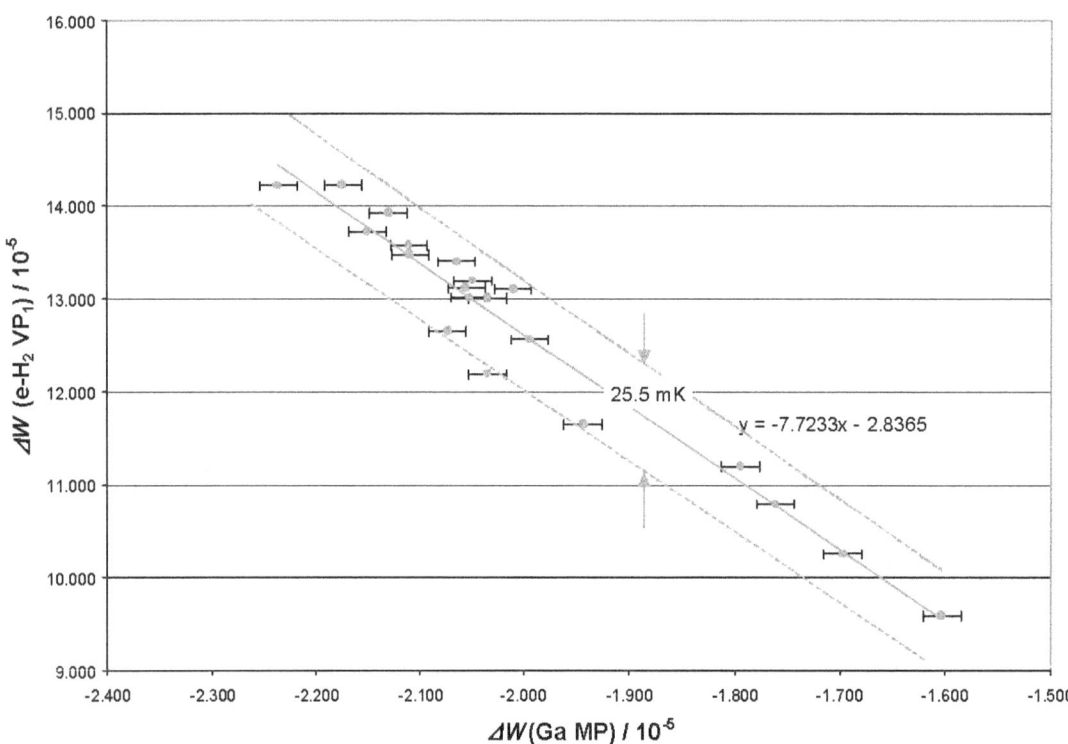

Figure 6e. The correlation plot for the deviations at the e-H_2 VP_1 versus the Ga MP. The central curve is a simple linear least-squares fit. The dotted lines are the symmetric bounds on the data about the linear fit with the size of the bound indicated as an equivalent temperature difference.

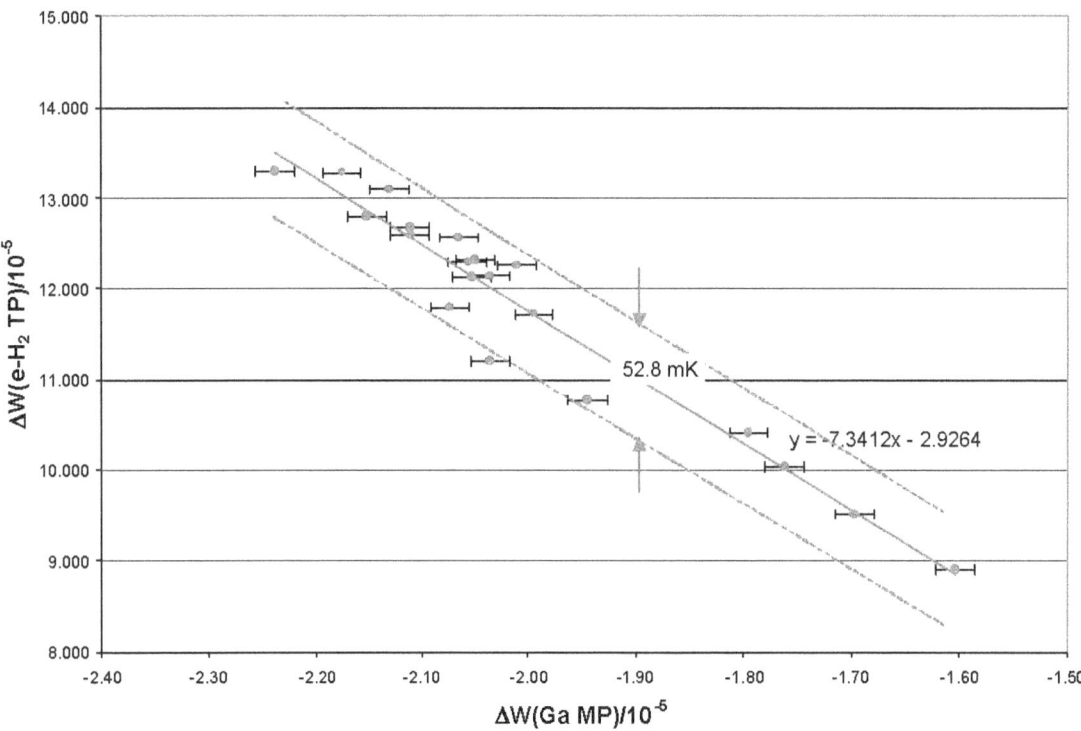

Figure 6f. The correlation plot for the deviations at the e-H_2 TP versus the Ga MP. The central curve is a simple linear least-squares fit. The dotted lines are the symmetric bounds on the data about the linear fit with the size of the bound indicated as an equivalent temperature difference.

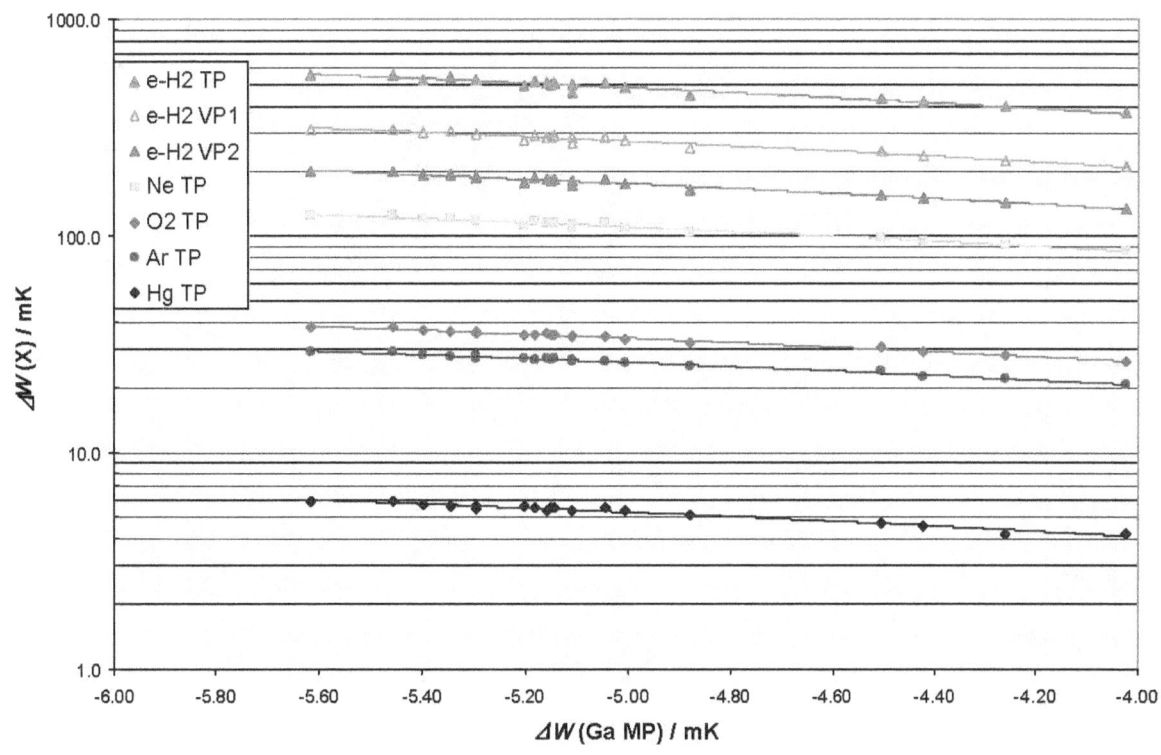

Figure 7. The deviation for the e-H_2 TP, e-H_2 VP_1, e-H_2 VP_2, Ne TP, O_2 TP, Ar TP, and Hg TP versus that for the Ga MP with the scales converted to equivalent temperature in mK.

Table 5. The statistical parameters for the linear fits shown in the Ga MP correlation plots of Figures 6a to 6f, and Figure 14.

	slope		intercept		residuals	
Fixed Point	m	σ_m	$\Delta_0 / 10^{-5}$	$\sigma_\Delta / 10^{-5}$	$\sigma_r / 10^{-5}$	σ_r / mK
In FP	5.46	0.16	0.61	0.32	0.112	0.28
Hg TP	-1.225	0.057	-0.31	0.11	0.040	0.10
Ar TP	-6.06	0.22	-0.80	0.44	0.155	0.36
O_2 TP	-7.32	0.22	-1.45	0.44	0.154	0.40
Ne TP	-8.01	0.31	-2.38	0.62	0.217	1.77
e-H_2 VP_2	-7.91	0.37	-2.64	0.74	0.257	3.42
e-H_2 VP_1	-7.72	0.42	-2.84	0.83	0.292	6.37
e-H_2 TP	-7.34	0.45	-2.93	0.91	0.317	13.20

3.1.2. Deviation Functions

The individual deviation functions $\Delta W(W(T_{90}))$ are shown in Figure 8 for the entire range of $W(T_{90})$ values spanned by the calibrations. The curves shown are the solutions to equation 7 and equation 8 for both batches of the SRM 1750 SPRTs. The complete set of fixed-point deviation data in the lower range is shown as a function of temperature in Figure 9.

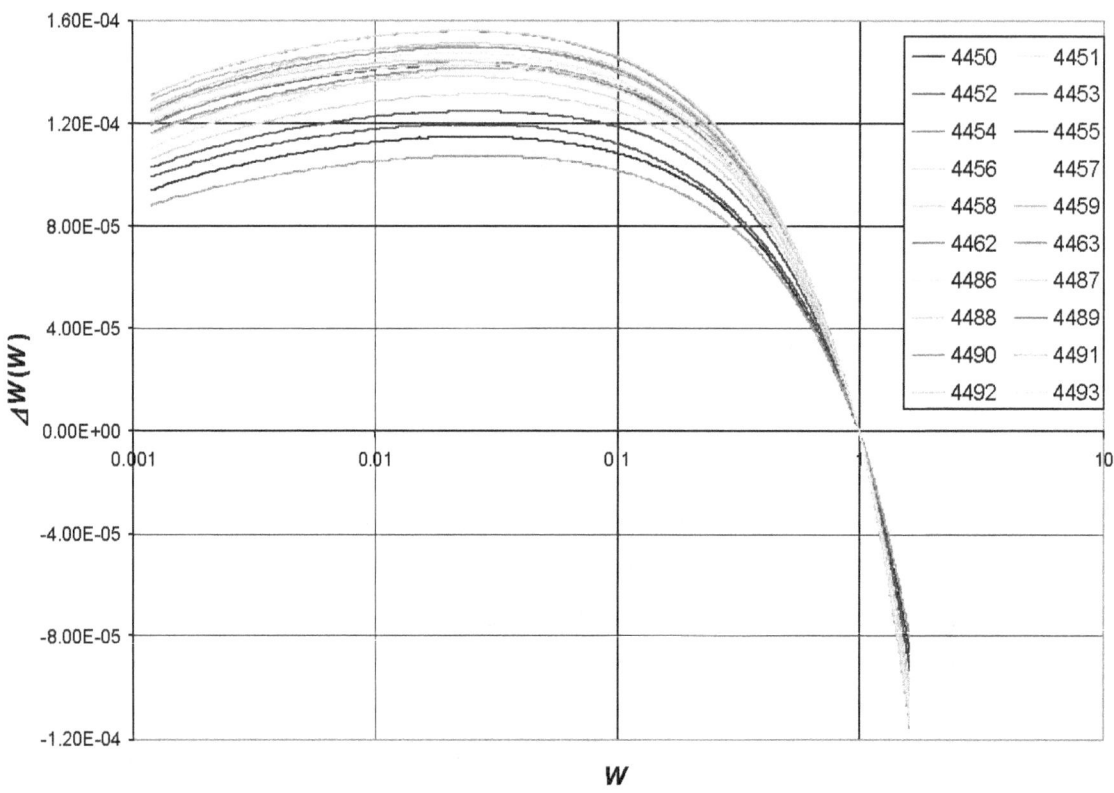

Figure 8. The individual deviation functions for the SRM 1750 shown as a function of W value.

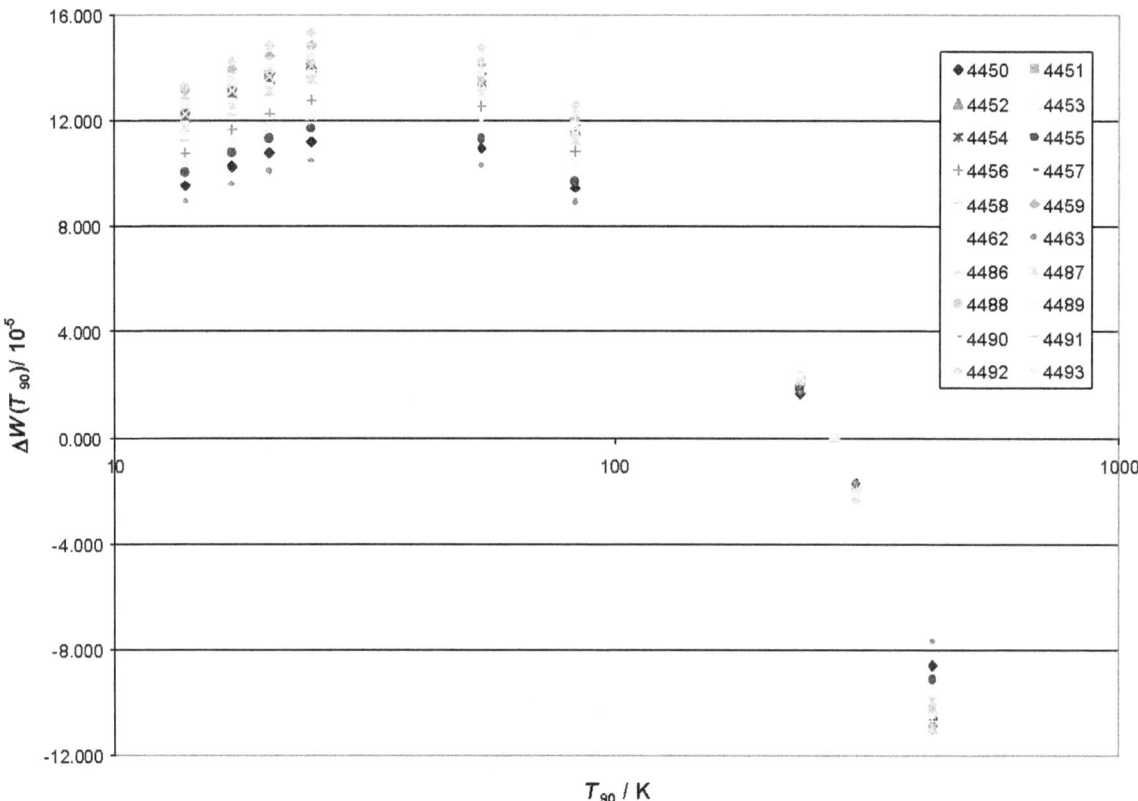

Figure 9. The individual deviations for the SRM 1750 shown versus temperature.

3.2. Uncertainties

The NIST assessment of uncertainties in the ITS-90 calibration of a thermometer involves the decomposition of uncertainty into Type A and Type B components. The Type A component, s, is a combined uncertainty from standard deviations of only those measurements under direct statistical process control. The Type B components, u_j, are the estimated standard uncertainties for each known component in the measurement process that are not directly measured.[31,32] In addition, uncertainties are described by a coverage factor, k, which allows an estimated uncertainty of the measurements to be expanded to a specific level of confidence. The resulting expanded uncertainty, when quoted with a coverage factor of $k = 2$ gives a level of confidence in the measurement of approximately 95 % when the degrees of freedom $v \geq 50$, as is consistent with international practice.

The uncertainties reported here do not include any estimates for: 1) the uncertainties that may exist in the possible different realizations of the ITS-90 between national standards laboratories in other countries and NIST; 2) the non-uniqueness of the ITS-90; 3) any effects that may be introduced by transportation of the thermometer between NIST and the user's laboratory; 4) long-term drift of the thermometer and; 5) any measurement uncertainties that exist in the user's laboratory.

3.2.1. Expanded Uncertainty

The Type A and Type B uncertainties at each of the SPRT calibration points are combined in quadrature and expanded by the coverage factor, k, to form the expanded uncertainty given by

$$U_T \equiv k\sqrt{s^2 + u_B^2} \ . \tag{10}$$

NIST expanded uncertainties are quoted with a coverage factor of $k=2$ unless stated otherwise. Table 6 lists the values of U_T in mK according to equation 10 for all of the ITS-90 calibration points relevant to SRM 1750. The values quoted from the Ar TP and higher are the latest estimates of uncertainty incorporating the most recent fixed-point material now in use at NIST. The estimates given here for some of the low-temperature comparison points are slightly larger than values given in reference [29] due to additional information on reference thermometer agreement obtained since the preparation of that earlier report.

Table 6. NIST expanded uncertainties for SRM 1750.

Calibration Point	T_{90} (K)	Expanded Uncertainty, U_T (mK)
e-H$_2$ TP[†]	13.8033	0.25
e-H$_2$ VP[†]	17.0	0.19
e-H$_2$ VP[†]	20.3	0.17
Ne TP[†]	24.5561	0.31
O$_2$ TP[†]	54.3584	0.14
Ar TP[†]	83.8058	0.21
Ar TP	83.8058	0.15
Hg TP	234.3156	0.20
H$_2$O TP	273.16	0.04
Ga MP	302.9146	0.04
In FP	429.7485	0.32

[†] Comparison Calibration

3.2.2. Total Calibration Uncertainty

The total calibration uncertainty at any temperature within any temperature sub-range is determined from the expanded uncertainties as propagated from each of the relevant defining fixed points. The uncertainty propagated from each defining fixed point is calculated by assuming the appropriate uncertainty at that fixed point but with no uncertainty at the other fixed points, and determining mathematically how that uncertainty propagates. The total uncertainty from a calibration is then determined in part by calculating the root-sum-square (RSS) uncertainty arising from all of the defining fixed points used in that calibration. Figures 10 and 11 show uncertainty propagation curves for the two ITS-90 temperature sub-ranges relevant to SRM 1750, using the uncertainties of the fixed points given in Table 6. The thick line represents the RSS uncertainty for the sub-ranges based on those uncertainties.

The propagated-uncertainty curve for the triple point of water (H$_2$O TP) (assuming 0.1 mK uncertainty at the H$_2$O TP), as shown in Figure 12, is the uncertainty incurred by the user, *not* an uncertainty in the NIST calibration. During the NIST calibration of a thermometer, any uncertainty from a measurement at the H$_2$O TP is incorporated into the definition of $W(T_{90})$. If the uncertainty made by the user at the H$_2$O TP is not 0.1 mK, a corrected propagated uncertainty at any temperature can be calculated by using the appropriate multiplicative factor.

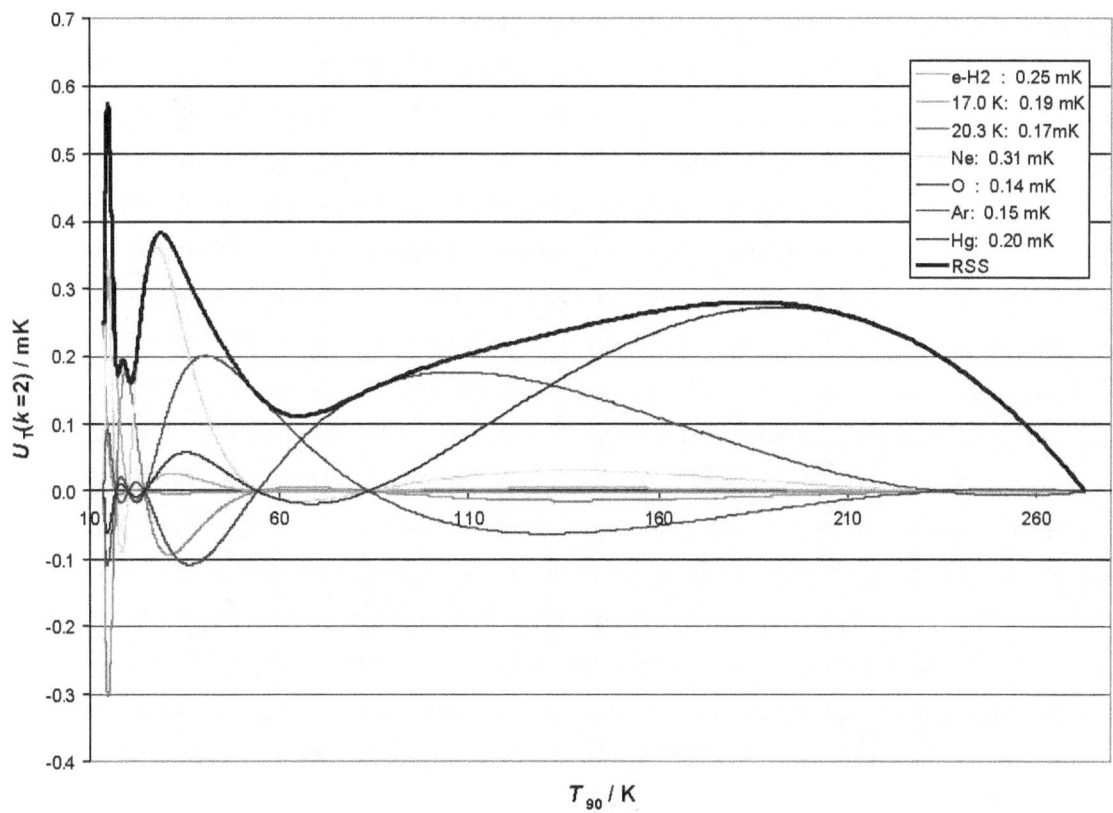

Figure 10. These curves show the uncertainties as propagated over the range 13.8 K to 273.16 K from the uncertainties for the calibration of SRM 1750 SPRTs (see Table 6) for each of the calibration fixed points within that range. The individual fixed-point uncertainty curves are added in quadrature to form the curve labeled "RSS" with only one (positive) sign shown in the figure which represents the total calibration uncertainty as reported by NIST.

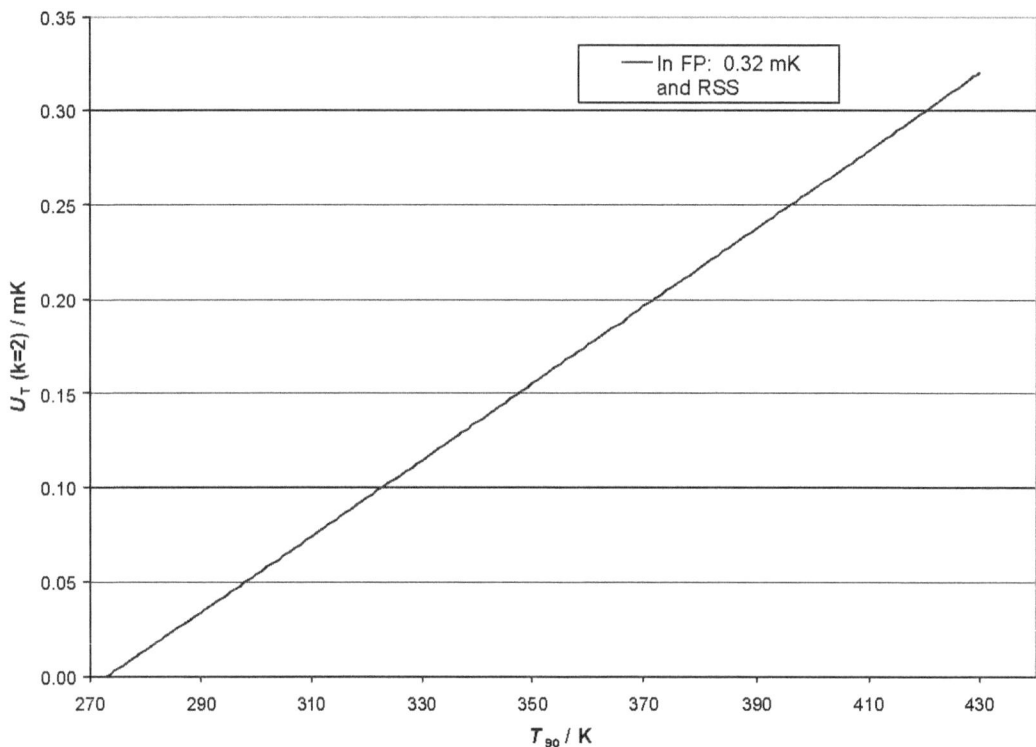

Figure 11. This curve shows the uncertainty as propagated over the range 273.15 K to 429.7485 K from the uncertainty for the calibration of SRM 1750 SPRTs (see Table 6) at the In FP. The curve is also equivalent to the "RSS," total calibration uncertainty in this range because the In FP is the only fixed point which contributes to the deviation equation in that sub-range.

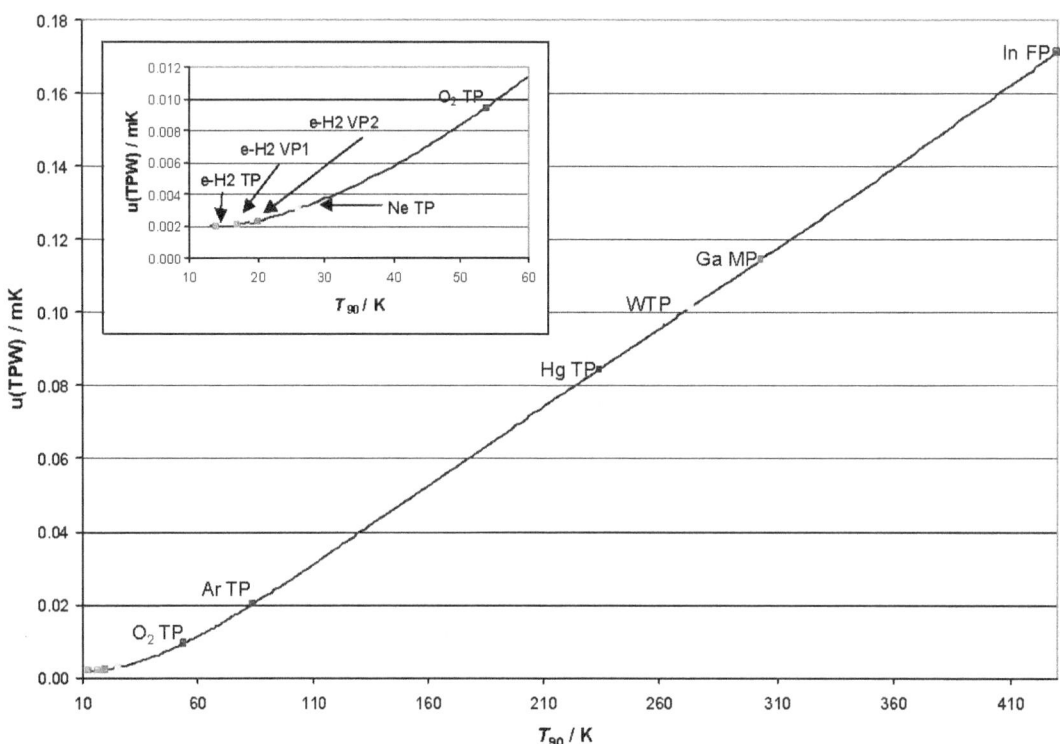

Figure 12. A separate uncertainty propagation curve for the triple point of water is shown to illustrate the calculation of the user's uncertainty if a user's measurement uncertainty at the triple point of water is 0.1 mK.

3.3. Supplementary Data

In addition to the data shown above for the 10 fixed-point temperatures, a number of supplementary data points were also obtained for SRM 1750. These data were obtained by comparison methods and include measurements of the resistance ratio at temperatures intermediate to some of the fixed-point temperatures (between 13.8 K and 83.8 K), as well as data below the lower temperature limit of the normal calibration (13.8033 K). The exact temperatures included for the supplementary data are somewhat variable, and the coverage is not entirely complete, depending on from which comparison batch the data were derived. Table 7 gives a breakdown of all the supplementary data for each SPRT in the SRM 1750 by serial number.

Table 7. Supplementary data point temperatures (in kelvin) for the 20 SPRTs of the SRM 1750.

Serial no.	Below 13.8 K	13.8 K to 17.0 K	17.0 K to 20.3 K	20.3 K to 24.56 K	24.56 K to 54.36 K		54.36 K to 83.8 K		
4450	4.2211	15.3946	18.6892	22.6002	—	—	61.7000	69.1006	77.3997
4451	4.2211	15.3946	18.6892	22.6002	—	39.4561	61.7000	69.1006	77.3997
4452	4.2211	15.3946	18.6892	22.6002	—	39.4561	61.7000	69.1006	77.3997
4453	4.2211	15.3946	18.6892	22.6002	—	39.4561	61.7000	69.1006	77.3997
4454	4.2211	15.3946	18.6892	22.6002	—	39.4561	61.7000	69.1006	77.3997
4455	4.2211	15.3946	18.6892	22.6002	—	39.4561	61.7000	69.1006	77.3997
4456	4.2211	15.3946	18.6892	22.6002	—	39.4561	61.7000	69.1006	77.3997
4457	4.2211	15.3946	18.6892	22.6002	—	39.4561	61.7000	69.1006	77.3997
4458	4.2211	15.3946	18.6892	22.6002	—	39.4561	61.7000	69.1006	77.3997
4459	4.2211	15.3946	18.6892	22.6002	—	39.4561	61.7000	69.1006	77.3997
4462	4.2209	15.3936	18.6887	22.5995	31.9969	39.4578	61.7001	69.0996	77.3985
4463	4.2209	15.3936	18.6887	22.5995	31.9969	39.4578	61.7001	69.0996	77.3985
4486	4.2209	15.3936	18.6887	22.5995	31.9969	39.4578	61.7001	69.0996	77.3985
4487	4.2209	15.3936	18.6887	22.5995	31.9969	39.4578	61.7001	69.0996	77.3985
4488	4.2212	15.3936	18.6887	22.5995	31.9969	39.4578	61.7001	69.0996	77.3985
4489	4.2209	15.3936	18.6887	22.5995	31.9969	39.4578	61.7001	69.0996	77.3985
4490	4.2209	15.3936	18.6887	22.5995	31.9969	39.4578	61.7001	69.0996	77.3985
4491	4.2209	15.3936	18.6887	22.5995	31.9969	39.4578	61.7001	69.0996	77.3985
4492	4.2209	15.3936	18.6887	22.5995	31.9969	39.4578	61.7001	69.0996	77.3985
4493	4.2209	15.3936	18.6887	22.5995	31.9969	39.4578	61.7001	69.0996	77.3985

Temperatures below 13.8 K were determined by the reference RIRT (s/n B-174) and check RIRT (s/n B-168) which were originally calibrated at NIST in 1995[33] according to the ^{4}He vapor pressure definition of the ITS-90. The estimated uncertainty on the comparison measurements of SRM 1750 SPRTs at 4.221 K was approximately 0.1 mK. All other supplementary data were obtained using the same reference thermometers and associated uncertainties as already described in sections 2.2 and 3.2.

3.3.1 Interpolation Characteristics (13.8 K to 83.8 K)

The accuracy of interpolation of temperatures between the fixed-point temperatures can be tested directly by performing comparison measurements of the SRM 1750 SPRTs at a set of such intermediate points. The deviation that is predicted for each SPRT on the basis of the reference thermometer temperature and the SPRT's individual deviation equations can then be compared with the actual measured deviation at those intermediate temperatures.

Figure 13a shows a plot from 13.8 K to 90 K of the difference between the predicted and the measured deviations for the first batch of the SRM 1750 SPRTs. The data include two additional check SPRTs (1842385 and 1812279) as well as two check RIRTs (B174 and B168). Figure 13b shows a plot from 13.8 K to 90 K of the difference between the predicted and the measured deviations for the second comparison batch of the SRM 1750 including two additional check SPRTs (1812282 and 1812284) as well as two check RIRTs (B174 and B168). The plots show all the comparison data at intermediate temperatures as well as at the Ar TP. At the fixed-point temperatures from the O_2 TP and below, however, only the check thermometers are shown because the SRM 1750 comparison data will by definition agree with the reference thermometer at these points. The comparison data at the Ar TP is with respect to the independent fixed-point data. The broad black lines represent the $k=2$ limits of uncertainty for the comparison measurements. These limits are somewhat larger than the limits on the actual calibration of the SRM 1750 SPRTs as shown in Figure 10 due to additional uncertainties imposed on the comparison data at 83.8 K and to an additional uncertainty on the reference thermometer's calibration at the Hg TP and Ar TP. These additional contributions to the comparison uncertainty affect only the accuracy of the comparison with the reference thermometer at the intermediate temperatures and do not affect the accuracy of the SRM 1750 calibrations.

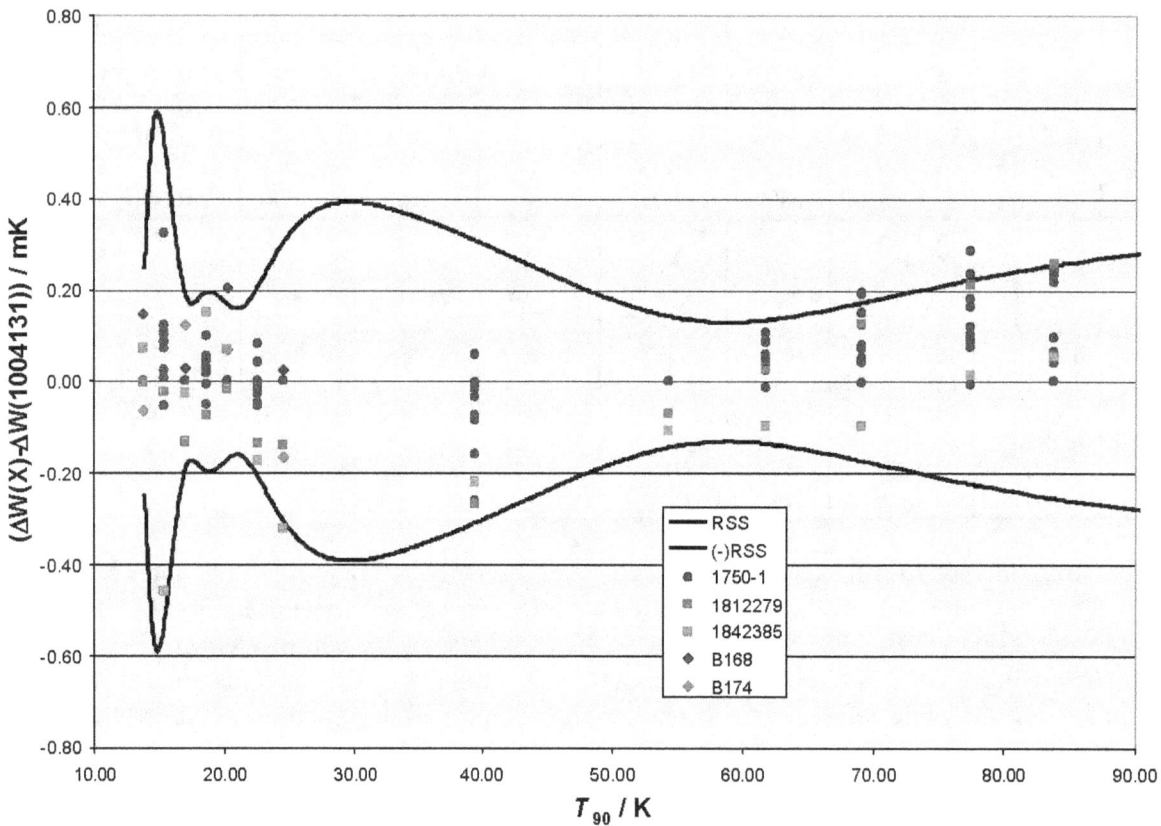

Figure 13a. The difference between the "as calibrated" deviation predicted by 10 individual SRM 1750 calibrations and the observed deviation according the NIST reference thermometer over the range 13.8 K to 83.8 K. The black line curves ("RSS") are the comparison uncertainty bounds ($k=2$) for the intermediate temperatures of the supplementary data from the first comparison batch of the SRM 1750.

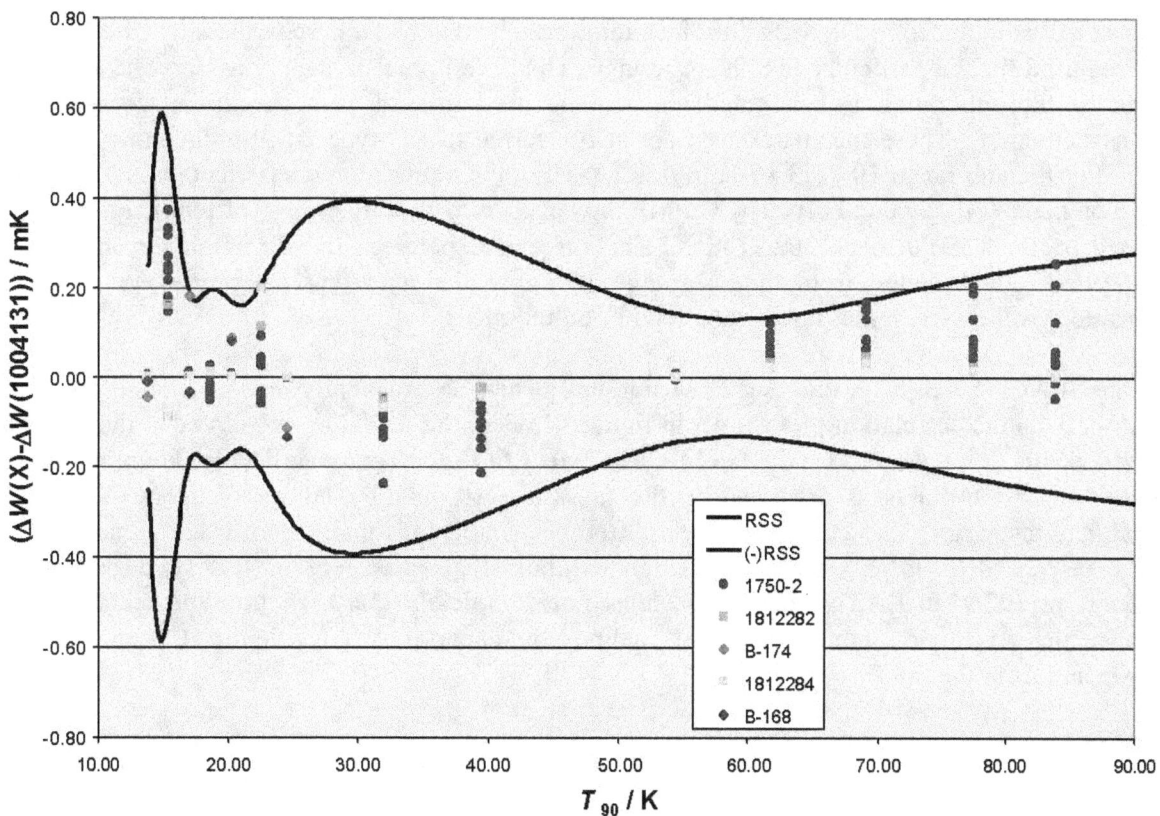

Figure 13b. The difference between the "as calibrated" deviation predicted by 10 individual SRM 1750 calibrations and the observed deviation according the NIST reference thermometer over the range 13.8 K to 83.8 K. The black line curves ("RSS") are the comparison uncertainty bounds ($k=2$) for the intermediate temperatures of the supplementary data from the second comparison batch of the SRM 1750.

3.3.2 Non-Uniqueness (302.9146 K)

The Ga MP is included in all NIST SPRT calibrations above 273.16 K regardless of whether or not it is required for the particular ITS-90 sub-range. This fixed point is then considered as redundant as far as the calibration is concerned and can be used instead as a checkpoint for sub-range inconsistencies.[38] These inconsistencies are also referred to as "type I non-uniqueness"[39] in the scale. For the sub-range 10 (273.15 K to 429.7485 K), the calibration coefficient a_{10} is calculated based on the $W(T_{InFP})$ value only. The Ga MP, however, is within this range of temperatures and is defined by the ITS-90 to be 302.9146 K. The agreement between the deviation for an SPRT at 302.9146 K as predicted from the sub-range 10 calibration and that determined via a Ga TP realization, is therefore a measure of this type I non-uniqueness.

For the SRM 1750 SPRTs, this degree of non-uniqueness is practically insignificant. This can be illustrated via the correlation plot shown in figure 14. Here the deviations observed at the In FP are plotted versus those observed at the Ga MP. The linear fit to the complete data is shown along with another fit to the points as predicted by the sub-range 10 (In FP) calibration alone. While it is possible to distinguish these two lines, their difference is not statistically significant. In addition, no SPRT in the SRM 1750 sample population exhibits a disagreement of this type greater than 0.12 mK at 302.9146 K. This disagreement is not completely due to non-uniqueness, however, because there is approximately 0.06 mK calibration uncertainty (see Figure 11) at 303 K as propagated from the In FP.

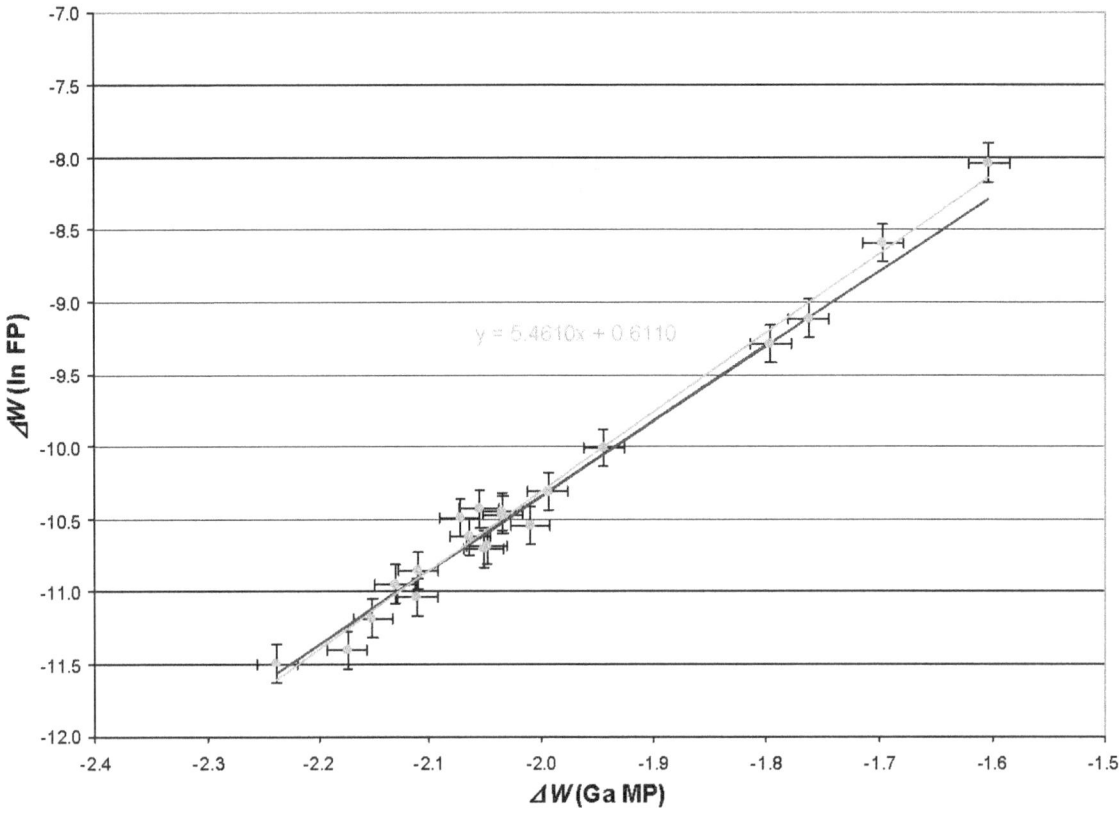

Figure 14. The correlation plot of the In FP and Ga MP as measured for the SRM 1750 population sample. The lighter line is a fit to the complete data (shown in equation form), and the darker line is a fit using the sub-range 10 calibration only (see text).

3.3.3. Residual Resistance Ratios (4.221 K)

Supplementary resistance ratio data at or very near the normal boiling point of liquid ^{4}He (^{4}He NBP) is of interest from the point of view of characterizing the purity of any nearly pure metal. Historically, the term "residual resistivity ratio," or "*RRR*,"[34] has been used to denote the following ratio of a metal's electrical resistivity $\rho(T)$,

$$RRR \equiv \frac{\rho(273.15 \text{ K})}{\rho(4.22 \text{ K})}, \tag{11}$$

however, the exact temperatures used have not always been precisely reported and/or entirely consistent. The *RRR* is usually calculated by taking the simple ratio of measured resistances with the assumption that the specific shape factors cancel in the ratio, leaving only a bulk resistivity ratio. The other inherent assumption made is that the temperature of 4.2 K is low enough to measure the true residual term in the resistivity, which is meant to be a purely temperature-independent term, dependent only on the impurity concentration of the individual sample. This assumption is true only if $R(4.22 \text{ K}) = R(0 \text{ K})$, which is generally satisfied for sufficiently high impurity concentrations. For very high purity material (low impurity concentration), however, such as the platinum in the SRM 1750, this is not necessarily the case. In fact, previous measurements at NIST[35] on platinum of comparable purity to the SRM 1750, have indicated that the true *RRR*, when derived from measurements close to 1 K, can be as much as 8 % higher than what would be otherwise inferred from the definition given in equation 11.

The raw measurements reported here at 4.221 K are parameterized in the same basic manner as is done for the other fixed-point temperatures, namely, $W(4.221 \text{ K})$ is calculated according to equation 1. However, the deviation $\Delta W(4.221 \text{ K})$ must be calculated in a different manner from that of other fixed points. This is because the ITS-90 reference function $W_r(T_{90})$ is essentially meaningless for temperatures below 13.8 K and returns an absurdly low value of approximately 1.25×10^{-27} at 4.221 K. Rather, for the purposes of the presentation of the data here, we adopt a "conventional value" for $W_r(T_{90}) \to W^*_r(4.221 \text{ K}) = 0.000348$. This conventional value is based on data[36] from the same capsule SPRT (s/n 217894) at NPL that was chosen as the basis of the ITS-90 reference function between 13.8033 K and 273.16 K.[37] Data are then presented in terms of the conventional deviations $\Delta W^*(4.221 \text{ K}) \equiv W(4.221 \text{ K}) - W^*_r(4.221 \text{ K}) = W(4.221 \text{ K}) - 0.000348$.

For the purposes of comparison of the SRM 1750 platinum to historical data on other samples of platinum found in the literature, we calculate the *RRR* starting from the definition of equation 11, and use the following approximation,

$$RRR \cong \frac{W_r(273.15 \text{ K})}{W(4.221 \text{ K})} = \frac{0.9999601}{W(4.221 \text{ K})}, \tag{12}$$

which accounts for the 10 mK difference between the ice point and triple point of water. This approximation assumes that $R(4.22 \text{ K}) = R(4.221 \text{ K}) = R(4.222 \text{ K})$, or that the difference is negligible. The resulting data for the SRM 1750 SPRTs from data obtained at 4.221 K are given in Table 8.

Table 8. The residual ratio data at 4.221 K as parameterized in three different ways (see text).

Serial no.	W(4.221 K)	ΔW^*(4.221 K)/ 10^{-5}	RRR
4450	0.00043249	8.449	2312
4451	0.00045960	11.160	2176
4452	0.00045415	10.615	2202
4453	0.00044288	9.488	2258
4454	0.00045323	10.523	2206
4455	0.00043541	8.741	2297
4456	0.00044387	9.587	2253
4457	0.00045544	10.744	2196
4458	0.00044384	9.589	2253
4459	0.00046285	11.485	2160
4462	0.00045432	10.632	2201
4463	0.00043037	8.237	2324
4486	0.00045452	10.652	2200
4487	0.00045028	10.228	2221
4488	0.00045774	10.974	2185
4489	0.00045819	11.019	2182
4490	0.00046376	11.576	2156
4491	0.00045960	11.160	2176
4492	0.00046329	11.529	2158
4493	0.00045717	10.917	2187

The degree of purity and absence of strain for each individual sample (SPRT) of SRM 1750 is then most easily accessed by inspection of the RRR values in Table 8. The highest RRR value of 2324 is only 7 % higher than the lowest value of 2156, which again is an indication of the relative uniformity of the SRM as a population. By contrast, the RRR for the NPL capsule SPRT (s/n 217894) mentioned above is approximately 2873, but this was perhaps the highest-purity sample of platinum wire of small (d<0.1 mm) diameter ever produced for any capsule-type SPRT. The expanded uncertainty in W(4.221 K) is 1.1×10^{-7} or approximately 0.025 % for the SRM 1750.

3.3.4. Low-Temperature Correlations

The data at 4.221 K are also suitable to illustrate another set of resistance ratio correlations that exist for the SRM 1750 platinum. These correlations are derived from the same fixed-point temperature deviations presented already in section 3.1, but now with the ΔW^*(4.221 K) values treated as the independent variable. These correlations are expected to be at least partially independent of those presented earlier using the Ga MP, as the mechanisms responsible for electron scattering in the low-temperature limit are dominated by impurities and other lattice defects. Figures 15a through 15f are the correlation plots of the deviations observed for SRM 1750 at the lowest six fixed points of e-H_2 TP, e-H_2 VP_1, e-H_2 VP_2, Ne TP, O_2 TP, and Ar TP respectively, to their observed deviations at 4.221 K.

Several general features can be seen in these low-temperature correlation plots. First, an overall linear trend is evident in all the plots, with some significant sample-to-sample departures from the linear fit. These sample dependent variations are bounded for the SRM 1750 population according to the limits shown on each of the figures, converted to their equivalent values in mK. These limits are set by $\pm 2\sigma_r$ for the residuals of the fits and are relatively constant for these lowest six fixed-point temperatures when expressed as pure (dimensionless) deviations: being approximately (0.3 to 0.2) $\times 10^{-5}$ for the temperature range studied. When expressed as equivalent temperature variations, the lowest temperatures appear to have wider limits for these variations, due to the rapidly decreasing value of dW/dT with decreasing temperature below about 30 K. When these sample variations are compared to the analogous variations shown in Figures 6, showing the correlations with the Ga MP, it is evident that these low-temperature fixed points are equally well or better correlated with the ^{4}He NBP deviations in the case of the e-H$_2$ TP, e-H$_2$ VP$_1$, and e-H$_2$ VP$_2$, but less so for the Ne TP and other higher-temperature fixed points. This should be expected to be the case simply due to the relative proximity of the respective temperatures, and to the increasing relative importance of the impurity scattering mechanism as temperature decreases. The linear fits shown in the Figure 15 series of correlation plots for the ^{4}He NBP are summarized in Table 9 in manner exactly analogous to that of Table 5 for the case of the Ga MP correlations.

Overall, the SRM 1750, as a sample population of platinum, exhibits correlations over the entire range of calibration as well as into a range of temperatures below 13.8 K. Deviation data from any one of the fixed-point temperatures for the SRM 1750 will in fact show a correlation with data from any of the other temperatures. The choice made in this analysis, to parameterize all such correlations in terms of only the two independent variables (Ga MP and ^{4}He NBP), is based partially on convenience and partially on the relatively low uncertainties that are placed on the data at those two fixed points. Moreover, the two temperatures (302.9 K and 4.221 K) are far enough apart that they serve to best separate the two predominate mechanisms, point defects and lattice vibrations, which are known to be responsible for the observed resistivity over the temperature range studied.

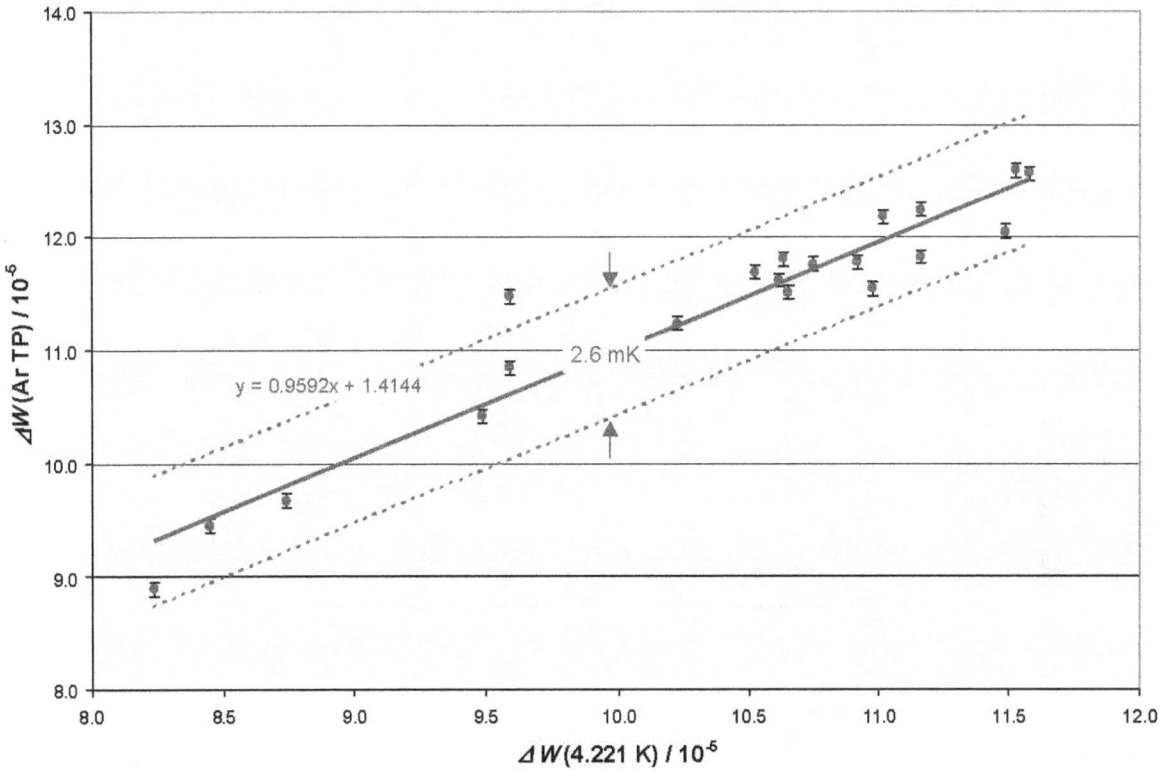

Figure 15a. The correlation plot for the deviations at the Ar TP versus the ^{4}He NBP. The central curve is a simple linear least-squares fit. The dotted lines are the symmetric bounds on the data about the linear fit with the size of the bound indicated as an equivalent temperature difference in mK.

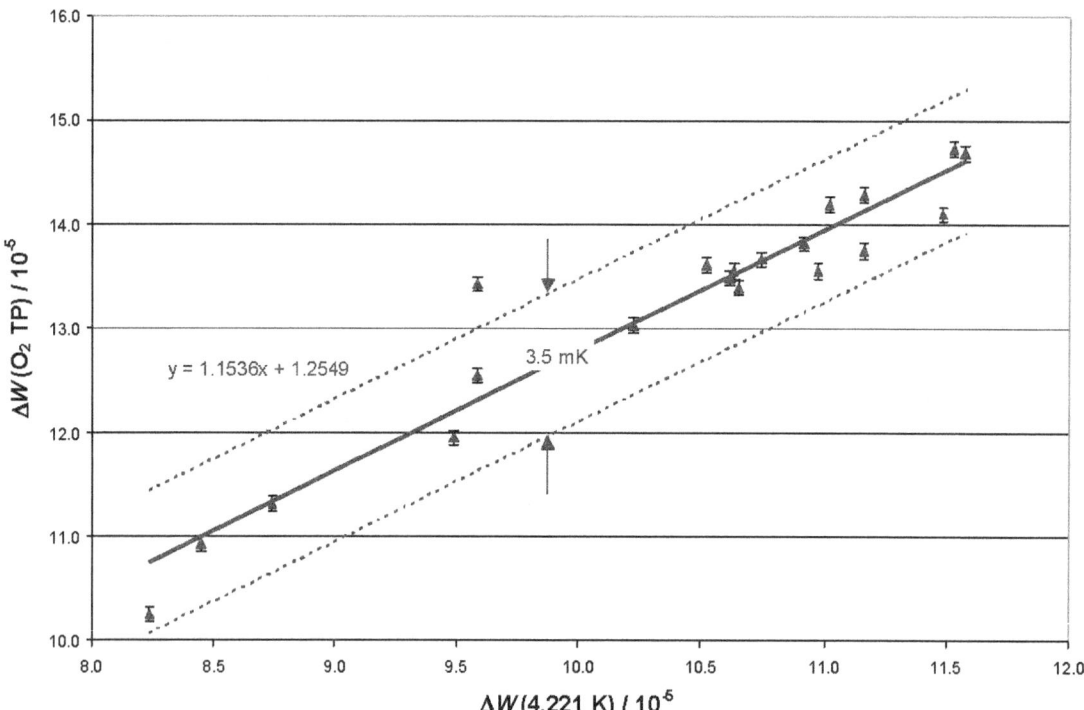

Figure 15b. The correlation plot for the deviations at the O_2 TP versus the ^{4}He NBP. The central curve is a simple linear least-squares fit. The dotted lines are the symmetric bounds on the data about the linear fit with the size of the bound indicated as an equivalent temperature difference in mK.

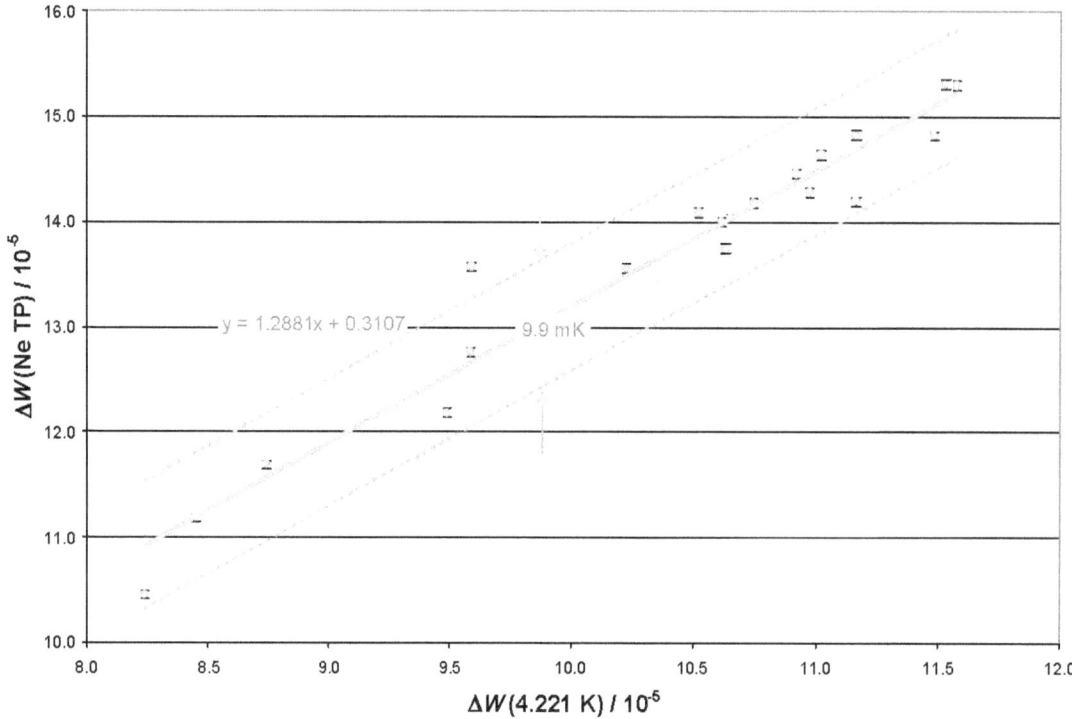

Figure 15c. The correlation plot for the deviations at the Ne TP versus the ^{4}He NBP. The central curve is a simple linear least-squares fit. The dotted lines are the symmetric bounds on the data about the linear fit with the size of the bound indicated as an equivalent temperature difference in mK.

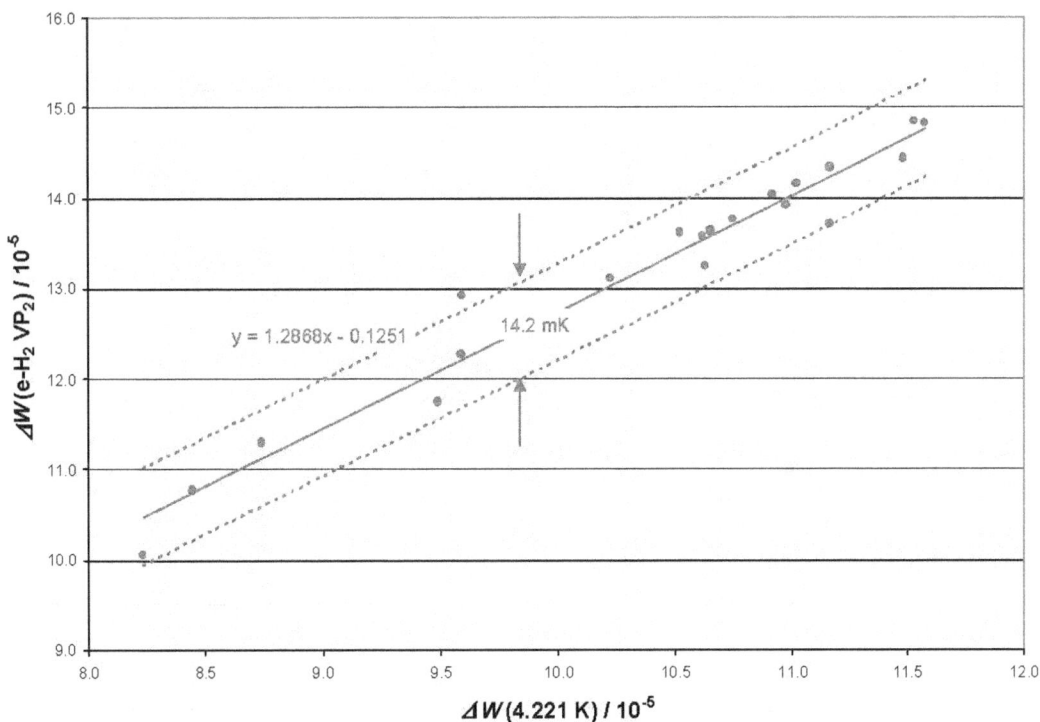

Figure 15d. The correlation plot for the deviations at the e-H_2 VP_2 (20.2714 K) versus the ^{4}He NBP. The central curve is a simple linear least-squares fit. The dotted lines are the symmetric bounds on the data about the linear fit with the size of the bound indicated as an equivalent temperature difference in mK.

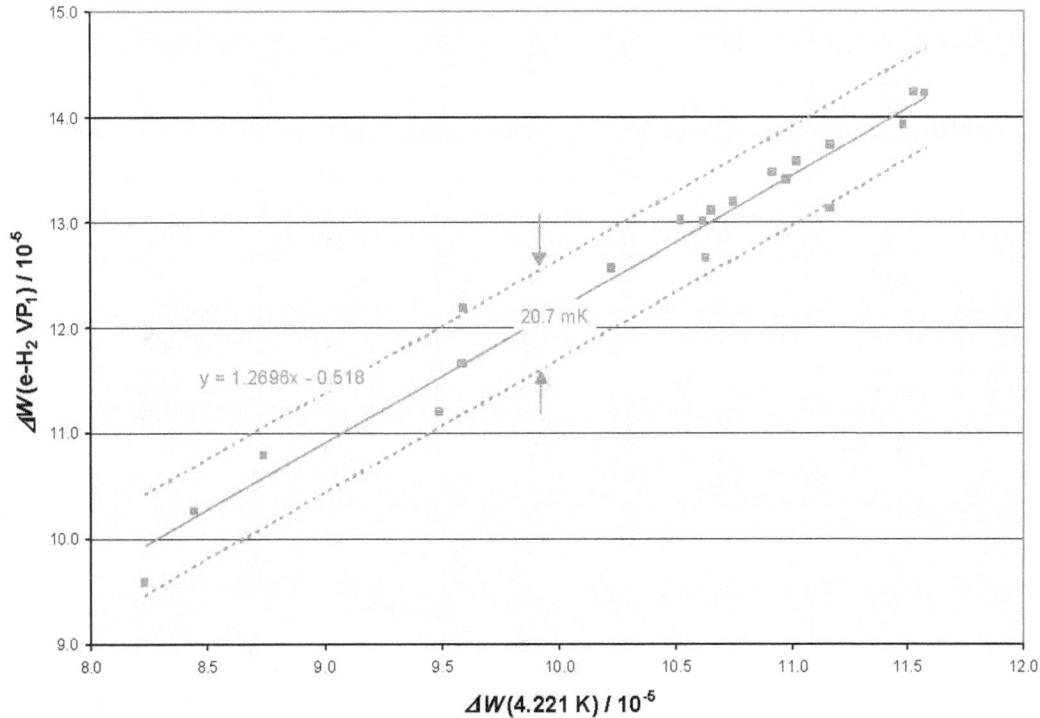

Figure 15e. The correlation plot for the deviations at the e-H_2 VP_1 (17.036 K) versus the ^{4}He NBP. The central curve is a simple linear least-squares fit. The dotted lines are the symmetric bounds on the data about the linear fit with the size of the bound indicated as an equivalent temperature difference in mK.

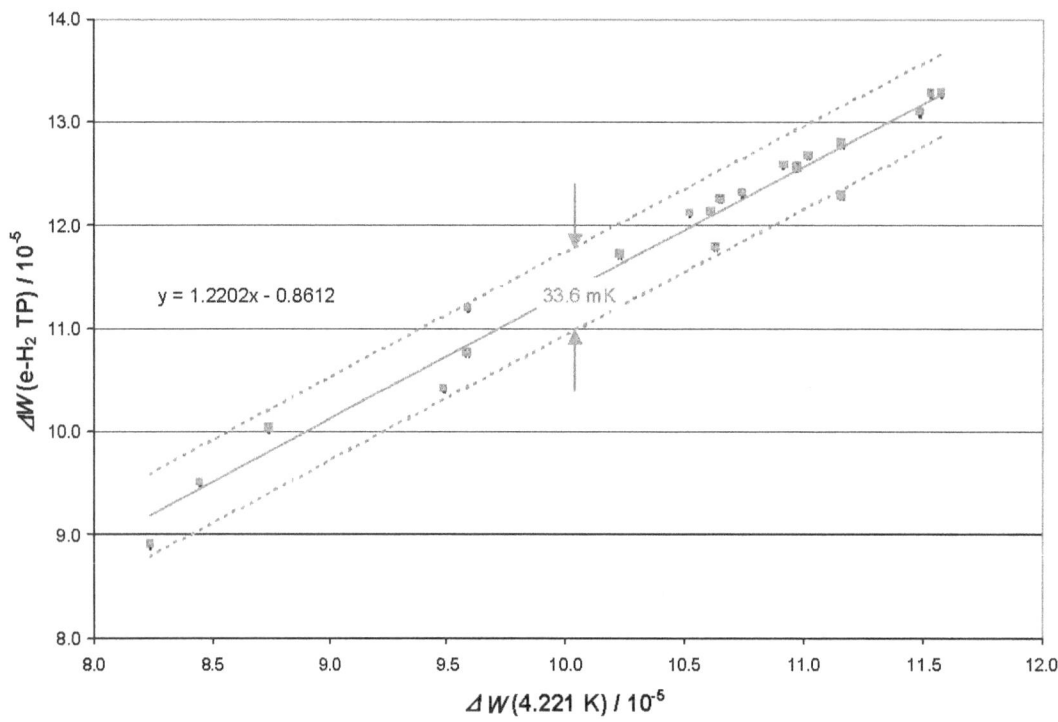

Figure 15f. The correlation plot for the deviations at the e-H$_2$ TP versus the ^{4}He NBP. The central curve is a simple linear least-squares fit. The dotted lines are the symmetric bounds on the data about the linear fit with the size of the bound indicated as an equivalent temperature difference in mK.

Table 9. The statistical parameters for the linear fits shown in the ^{4}He NBP correlation plots of Figures 15a to 15f.

	slope		intercept		Residuals	
Fixed Point	m	σ_m	$\Delta_0 / 10^{-5}$	$\sigma_\Delta / 10^{-5}$	$\sigma_r / 10^{-5}$	σ_r / mK
Ar TP	0.959	0.066	1.41	0.69	0.287	0.66
O$_2$ TP	1.154	0.079	1.26	0.83	0.342	0.88
Ne TP	1.288	0.070	0.31	0.73	0.303	2.46
e-H$_2$ VP$_2$	1.287	0.062	-0.13	0.65	0.267	3.55
e-H$_2$ VP$_1$	1.270	0.055	-0.52	0.57	0.237	5.18
e-H$_2$ TP	1.220	0.047	-0.86	0.49	0.202	8.40

4. SUMMARY

The SRM 1750 is both a sample population of high-purity platinum wire elements as well as a collection of highly accurate reference thermometers calibrated according to the definitions of the ITS-90. The population has resistance characteristics which are both highly uniform and highly correlated over the entire range of temperatures studied. The SRM 1750 W(Hg TP) and W(Ga MP) values significantly exceed the ITS-90 criterion for acceptance as a defining instrument of the scale.

The starting material for the platinum wire elements is know to be 99.999 % pure, but the actual purity obtained in the final SPRT units is expected to be slightly lower. Based on the residual resistivity ratio data, the actual impurity concentrations within the platinum wire elements probably vary only on the order of 7 % throughout the SRM 1750 sample population. The lowest *RRR* observed for the SRM 1750 was only 25 % smaller than that of the reference function SPRT.

Each individual unit of the SRM is fully certified by NIST as suitable to serve as a defining instrument of the ITS-90 and is calibrated at the highest accuracy level available at NIST. All certified values are determined by reference to ITS-90 defined fixed points and the ITS-90 SPRT reference function. The values which are NIST certified are the deviation function coefficients of each individual SPRT for sub-range 1 and sub-range 10 on the ITS-90. These coefficients form the SPRT calibration for use at either 1 mA or 2 mA excitation current or by extrapolation to 0 mA.

Each unit of SRM 1750 is provided with a glass adapter probe for use in suitable fixed-point cells at temperatures between 83.8 K and 430 K. The user can then periodically check the SPRT in a H_2O TP cell using the same probe adapter in which it was originally calibrated, in order to maintain the highest quality assurance.

5. REFERENCES

1. H. Preston-Thomas, "The International Temperature Scale of 1990 (ITS-90)," *Metrologia* 27, 3 (1990); ibid, 107 (1990).
2. B. W. Mangum and G. T. Furukawa, *Guidelines for Realizing The International Temperature Scale of 1990 (ITS-90)*, NIST Technical Note 1265, (U.S. Government Printing Office, Washington, DC, 1990).
3. Sigmund Cohn Corporation, Mount Vernon, NY, Platinum bar lots 240-203 and 240-206.
4. J. L. Riddle, G. T. Furukawa, and H. H. Plumb, *Platinum Resistance Thermometry*, NBS Monograph 126, p. 8, (U.S. Government Printing Office, Washington, DC, April 1973).
5. B. W. Mangum, *Platinum Resistance Thermometer Calibrations*, NBS Special Publication 250-22, pp. 14-19 (U.S. Government Printing Office, Washington, DC, October 1987).
6. G. F. Strouse, "NIST implementation and realization of ITS-90 over the range 83 K to 1235 K," in *Temperature: Its Measurement and Control in Science and Industry*, Vol. 6, p169, J. F. Schooley, ed., American Institute of Physics, New York (1992).
7. W. L. Tew, G. F. Strouse, C. W. Meyer, and G. T. Furukawa, "Recent Advances in the Realization and Dissemination of the ITS-90 Below 83.8058 K at NIST," *Advances in Cryogenic Engineering*, Vol. 43B, P. Kittel, editor (Plenum Press, New York, 1998).
8. W. L. Tew and B. W. Mangum, "New Procedures and Capabilities for the Calibration of Cryogenic Resistance Thermometers at NIST," in *Advances in Cryogenic Engineering*, Vol. 39B, p. 1019, P. Kittel, editor, (Plenum Press, New York, 1994).
9. J. L. Riddle et al., *op. cit.*, p. 29.
10. F. Pavase and G. Molinar, "Modern Gas-Based Temperature and Pressure Measurements," p. 92, Plenum Press, New York (1992).
11. J. L. Riddle et al., *op. cit.*, p. 13.
12. B. W. Mangum and G. T. Furukawa, *op. cit.*, p. 25.
13. B. W. Mangum and G. T. Furukawa, *op. cit.*, pp. 42-47.
14. G. F. Strouse, NIST Special Publication 260-132.
15. G. F. Strouse, "NIST Realization of The Gallium Triple Point," submitted for publication, *Proceedings of TEMPMEKO '99*, Delft, The Netherlands, June 1999.
16. Jarrett Instrument Co., Inc., Wheaton, MD, USA. All the H_2O TP cells listed were produced in the Wheaton facility and are less than 6 years old. As of 1997, the Jarrett name and production process has been owned by Isothermal Technology, Ltd., Pine Grove, Southport, England, U.K.
17. G. T. Furukawa, B. W. Mangum, and G. F. Strouse, "Effects of different methods of preparation of ice mantles of triple point of water cells...," *Metrologia*, Vol. 34, pp. 215-233 (1997).
18. G.T. Furukawa, "Realization of the mercury triple point," in *Temperature: Its Measurement and Control in Science and Industry*, Vol. 6, p. 281, J. F. Schooley, ed., American Institute of Physics, New York (1992).
19. G. T. Furukawa, J. L. Riddle, W. R. Bigge, E. R. Pfeiffer, *Application of Some Metal SRM's as Thermometric Fixed Points*, National Bureau of Standards Special Publication 260-77, pp. 87-102, U.S. Government Printing Office, Washington, DC (1982).
20. G. T. Furukawa, "Argon triple point apparatus with multiple thermometer wells," in *Temperature: Its Measurement and Control in Science and Industry*, Vol. 6, p. 265, J. F. Schooley, ed., American Institute of Physics, New York (1992).
21. Airco Rare and Specialty Gas, Riverton, NJ, USA.
22. W. L. Tew *et al.* (1998), *op. cit.*

23. C. W. Meyer and M. L. Reilly, "Realization of the ITS-90 Triple Points From 13,80 K to 83,8 K at NIST," *Proceedings of the IMEKO International Seminar on Low Temperature Thermometry and Dynamic Temperature Measurement*, pp. L-110 – L-115, IMEKO, Wroclaw, Poland, (September 1997).
24. W. L. Tew, "Sealed-cell devices for the realization of the triple point of neon at the National Institute of Standards and Technology," *Proceedings of TEMPMEKO '96*, P. Marcarino, editor, p. 81, Levrotto and Bella, Torino, Italy (1997).
25. G. T. Furukawa, "Reproducibility of the triple point of argon in sealed transportable cells," in *Temperature: Its Measurement and Control in Science and Industry*, Vol. 5, p. 239, J. F. Schooley, ed., American Institute of Physics, New York (1982).
26. G. T. Furukawa, "The triple point of oxygen in sealed transportable cells," *J. Res. Nat. Bur. Stand.* (U.S.), 91, 255 (1986).
27. C. W. Meyer, G. F. Strouse, and W. L. Tew, "Non-Uniqueness of the ITS-90 from 13.8033 K to 24.5561 K," submitted for publication, *Proceedings of TEMPMEKO '99*, Delft, The Netherlands, June 1999.
28. H. Van Dijk, "On the Use of Platinum Thermometers for Thermometry Below 90 K...," in *Temperature: Its Measurement and Control in Science and Industry*, Vol. 3, p. 365, H. Plumb, ed., Reinhold Publishing Corp., New York (1962).
29. W. L. Tew, G. F. Strouse, and C. W. Meyer, *A Revised Assessment of Calibration Uncertainties for Capsule Type Standard Platinum and Rhodium-Iron Resistance Thermometers*, NISTIR 6138 (U.S. Government Printing Office, Washington, DC, April 1998).
30. G. F. Strouse and W. L. Tew, *Assessment of Uncertainties of Calibration of Resistance Thermometers at the National Institute of Standards and Technology*, NISTIR 5319, (U.S. Government Printing Office, Washington, DC, January 1994).
31. B. N. Taylor and C. E. Kuyatt, *Guidelines for Evaluating and Expressing the Uncertainty of NIST Measurement Results*, NIST Technical Note 1297, 1994 Edition, 20 pp. (September 1994).
32. ISO, *Guide to the Expression of Uncertainty in Measurement*, International Organization for Standards, Geneva, Switzerland (1993).
33. C. W. Meyer and M. L. Reilly, "Realization of the ITS-90 at the NIST in the range 0,65 K to 5,0 K using ^{3}He and ^{4}He vapour pressure thermometry," *Metrologia*, Vol. 33, p. 383, (1996).
34. F. R. Fickett, "Electrical Properties," in *Materials at low temperatures*, R. P. Reed and A. F. Clark, eds., pp. 163-201, American Society for Metals, Metals Park, OH (1983).
35. E. R. Pfeiffer, NIST, unpublished data on SPRT 1812279, Gaithersburg, MD (1992).
36. R. L. Rusby, private communication, NPL, Teddington, England, UK (Jan. 13, 1999).
37. R. C. Kemp, "The Reference Function for PRT Interpolation Between 13.8033 K and 273.16 K in the ITS-90," *Metrologia*, Vol. 28, pp. 327-332, (1991).
38. G. F. Strouse, "NIST assessment of ITS-90 non-uniqueness for 25.5 Ω SPRTs at gallium, indium, and cadmium fixed points," in *Temperature: Its Measurement and Control in Science and Industry,*" Vol. 6, p. 175, J. F. Schooley, ed., American Institute of Physics, New York (1992).
39. B. W. Mangum et al., "On the International Temperature Scale of 1990 (ITS-90), Part I: Some Definitions," *Metrologia*, Vol. 34, pp. 427-429 (1997).

www.ingramcontent.com/pod-product-compliance
Lightning Source LLC
Chambersburg PA
CBHW081804170526
45167CB00008B/3316